P9-DMC-344

Nuclear Crisis Management

Cornell Studies in Security Affairs

edited by Robert J. Art *and* Robert Jervis

Strategic Nuclear Targeting, edited by Desmond Ball and Jeffrey Richelson
Citizens and Soldiers: The Dilemmas of Military Service, by Eliot A. Cohen
Great Power Politics and the Struggle over Austria, 1945–1955, by Audrey Kurth Cronin
The Wrong War: American Policy and the Dimensions of the Korean Conflict, 1950–1953, by Rosemary Foot
The Soviet Union and the Failure of Collective Security, 1934–1938, by Jiri Hochman
The Warsaw Pact: Alliance in Transition? edited by David Holloway and Jane M. O. Sharp
The Illogic of American Nuclear Strategy, by Robert Jervis
Nuclear Crisis Management: A Dangerous Illusion, by Richard Ned Lebow
The Nuclear Future, by Michael Mandelbaum
Conventional Deterrence, by John J. Mearsheimer
The Sources of Military Doctrine: France, Britain, and Germany between the World Wars, by Barry R. Posen
The Ideology of the Offensive: Military Decision Making and the Disasters of 1914, by Jack Snyder
The Militarization of Space: U.S. Policy, 1945–1984, by Paul B. Stares
Making the Alliance Work: The United States and Western Europe, by Gregory F. Treverton
The Ultimate Enemy: British Intelligence and Nazi Germany, 1933–1939, by Wesley K. Wark

Nuclear Crisis Management

A DANGEROUS ILLUSION

RICHARD NED LEBOW

Cornell University Press

ITHACA AND LONDON

First published 1987 by Cornell University Press.

International Standard Book Number 0-8014-1989-1
Library of Congress Catalog Card Number 86-16767
Printed in the United States of America.
Librarians: Library of Congress cataloging information appears on the last page of the book.

The paper in this book is acid-free and meets the guidelines for permanence and durability of the Committee on Production Guidelines for Book Longevity of the Council on Library Resources.

Contents

To David and his generation.
May they never know war.

Preface

Two strands of my life come together in this book. The first, the intellectual, involves eight years devoted to researching and writing about crises. The second, the professional, consists of five years of government service at the naval and national war colleges and the Central Intelligence Agency, much of that time concerned with the prediction and management of crisis.

A previous book, *Between Peace and War,* was a theoretical and historical study of crisis. It made no mention of superpower confrontations involving nuclear alerts other than that in Cuba. I was nevertheless convinced that many of my findings about the origins and dynamics of crisis were relevant to present-day Soviet-American relations.

My years in government agencies convinced me of the pressing need for a book that spells out the implications of historical experience for the prevention and management of crises. The National War College, where I was professor of strategy for two years, has a banal curriculum that stresses the political-military orthodoxy of the day. It is nevertheless a wonderful window into the world of national security policy. Several times a week top officials from the departments of State and Defense and the White House make the short trip to Fort McNair to give off-the-record lectures. Sometimes I had the opportunity to confer with them privately. I also wrote and supervised crisis management exercises in which some of these officials took part. The "students" I taught, officers and civil servants with an average of twenty years of service, had themselves often been key players in past crises, military operations, or arms control negotiations.

The lectures, classes, and war games made me painfully aware of the stereotyped conceptions that so many of these officials held about the Soviet Union and of how they were committed to traditional Cold War strategies for coping with the threat the Russians posed. The lack of political empathy shown by these officials, their parochial view of the world, and their unyielding emphasis on military capability and the need to demonstrate resolve struck me as inappropriate to a world of nuclear weapons. Perhaps this outlook explains why the teams that played the crisis exercises I designed so often blundered into war.

My experience was no different at the Central Intelligence Agency, where I spent a year as scholar-in-residence. Most of the time I worked in what was then called the Strategic Evaluation Center, where I had ample opportunity to study the assumptions and methods used by Agency analysts to evaluate Soviet strategic capability and intentions. Strategic assessment in the CIA—and even more in the Defense Intelligence Agency and the military services—reflects a narrow fixation on relative military capability. It assumes that Soviet aggressiveness is largely a function of the balance of military power. This is hardly surprising, for the strategic bureaucracy within the CIA is dominated by former military officers who are technically sophisticated but whose political outlook has been shaped by their service backgrounds. The colonels' view of the world, dominated by the operational characteristics of military forces, is rarely challenged by higher officials who, in many cases, share this apolitical outlook themselves.

Official U.S. thinking has remained rigid and doctrinaire at a time when international relations have become more complex and the consequences of miscalculation incalculably tragic. I have no illusion that scholars can change this disturbing state of affairs, but we do have a responsibility to present more enlightened and constructive perspectives. With this end in mind, I began to work on a critical evaluation of the assumptions about nuclear weapons, the Soviet Union, and crisis management which have shaped U.S. security policy. This book deals with American understandings of the place of crisis management in superpower strategic relations. A second volume will treat the underlying causes of U.S.-Soviet strategic rivalry.

This book was greatly facilitated by a generous grant from the Carnegie Corporation of New York. I am grateful to David Hamburg, president, and Fritz Mosher, program officer, for the interest they have shown, and continue to show, in my work. I am also indebted to my neighbor and colleague Kurt Gottfried, who organized and directed an interdisciplinary research project on crisis stability cosponsored by the Peace Studies Program of Cornell University and the American Academy of Arts and Sciences. That project brought together civilian and military authorities on strategic command and control for a series of seminars in Ithaca and Cambridge. My understanding of this subject profited enormously from the seminars and from conversations with the participants. This too was made possible by the largess of the Carnegie Corporation, the principal source of funding for the project.

Authors may be solely responsible for their books, but they profit from the publications of others, from dialogue with colleagues, and, ultimately, from readers' comments. In this regard I have been remarkably fortunate. I have learned a lot from existing studies of command and control. Colleagues were forthcoming with assistance at every stage of research and writing. Janice Gross Stein, Desmond Ball, and Matthew Evangelista were particularly helpful; they supplied references, good advice, and penetrating critiques almost on demand. Michael MccGwire and David Holloway were kind enough to share with me their extensive substantive and bibliographical knowledge of the Soviet Union. Alexander George, Irving Janis, Janice Gross Stein, and Robert Jervis read the book in manuscript and made numerous valuable suggestions for improving it. So did Roger Haydon, my editor at Cornell University Press.

I also acknowledge the help of John Garafano, my research assistant; David Cohen, who came to my rescue whenever I called him in a panic, convinced that I had lost part of my manuscript in the word processor; and Peace Studies administrative assistants Kathleen O'Neill and Helen Shelley and secretary Christine Hammon. Above all, I am grateful to my family for their support and tolerance. "Leave daddy alone, he's working on his book," was a command that rang out far too often in the course of the last year.

Richard Ned Lebow

Ithaca, New York, and
Wellington, New Zealand

Abbreviations

ASAT	Anti-Satellite Weapon
BMD	Ballistic Missile Defense
BMEWS	Ballistic Missile Early Warning System
C^3	Command, Control, Communication
C^3I	Command, Control, Communication, and Intelligence
DefCon	Defense Condition (U.S. military alert status)
DEW	Distant Early Warning
DSCS	Defense Satellite Communication Systems
DSP	Defense Support Program
EAM	Emergency Action Message
EMP	Electro-Magnetic Pulse
Ex Com	Executive Committee of the National Security Council (body convened by President Kennedy to help him manage the Cuban missile crisis)
GPS	Global Ranging Positioning System
GWEN	Ground Wave Emergency Network
ICBM	Inter-Continental Ballistic Missile
JSTPS	Joint Strategic Target Planning Staff
KGB	Komitet Gosudarstvenney Bezopasnosti; Committee for State Security (Soviet)
LOW	Launch On Warning
LUA	Launch Under Attack
MGT	Mobile Ground Terminal
MILSTAR	Military Strategic/Tactical and Relay Satellite
MIRV	Multiple Independently Targeted Re-Entry Vehicle
NATO	North Atlantic Treaty Organization
NAVSTAR	Navigation Satellite Timing and Ranging (later became GPS)

NCA	National Command Authority
NEACP	National Emergency Airborne Command Post
NORAD	North American Air Defense
PAL	Permissive Action Link
PARCS	Perimeter Acquisition Radar Attack Characterization System
PAVE PAWS	Precision Acquisition of Vehicle Entry Phased Array Warning System
SAC	Strategic Air Command
SACLANT	Supreme Allied Commander, Atlantic
SALT	Strategic Arms Limitation Treaty (talks)
SLBM	Submarine Launched Ballistic Missile
SSBN	Ballistic Missile Submarine, Nuclear Powered
TACAMO	Take Charge And Move Out (U.S. naval airborne communication relay system)

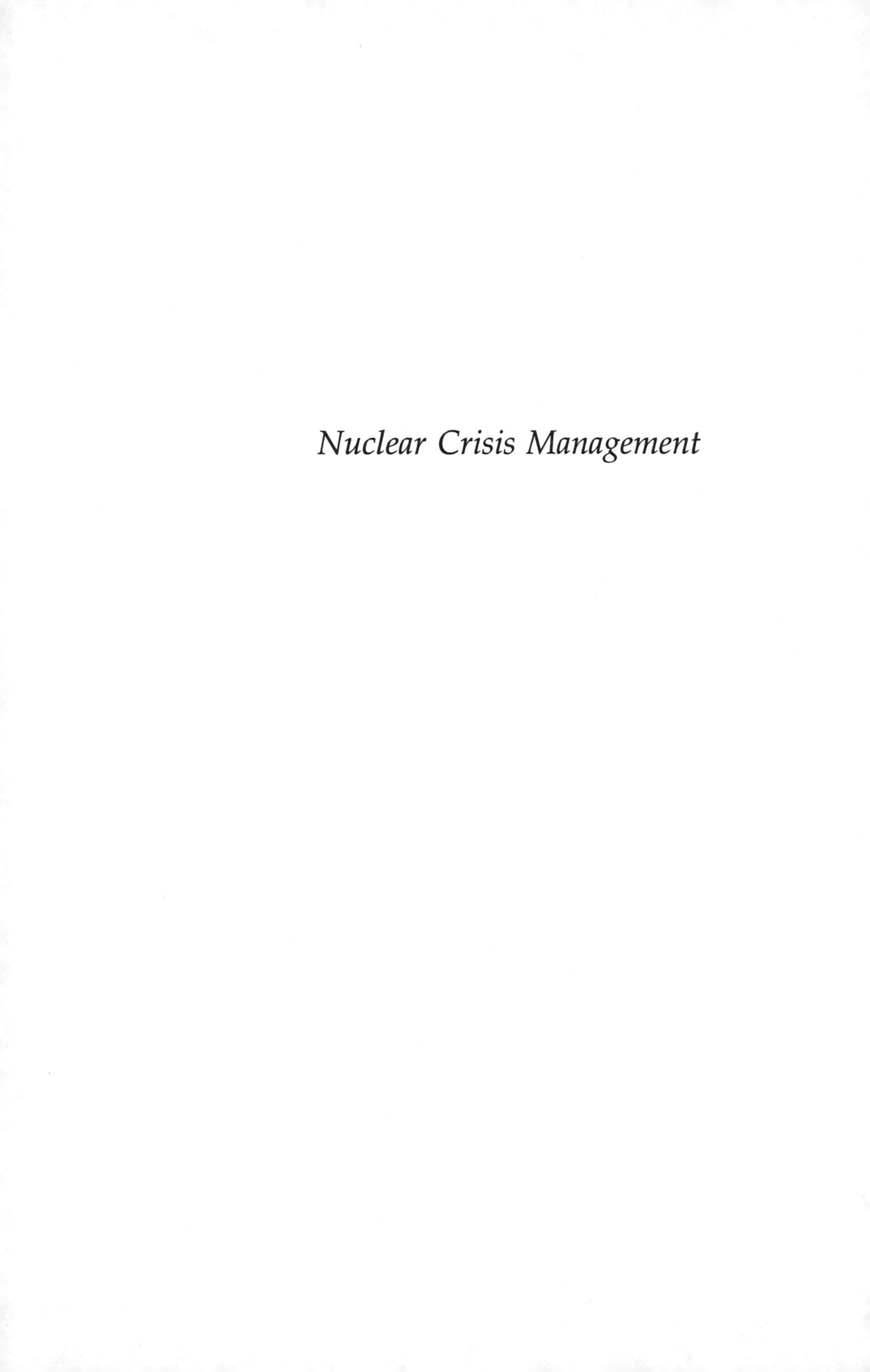

Nuclear Crisis Management

[1]

A Dangerous Illusion

After the Cuban missile crisis had been resolved, President Kennedy confessed that the odds of the Soviets going to war seemed to him at the time "somewhere between one out of three and even." "Our escape," he told Theodore Sorensen, "seems awesome."[1] Kennedy's judgment has remained a matter of controversy among scholars and members of the "Ex Com," the informal working group the president assembled to help him manage the crisis. Robert McNamara and McGeorge Bundy contend that it may have been a reasonably accurate estimate of the risks at the time.[2] Dean Acheson and Paul Nitze, by contrast, argued that the president greatly exaggerated the prospect of war.[3] Even if he did, Kennedy's judgment nevertheless reflects an undisputed political reality: the Cuban missile crisis was, relatively speaking, the closest the superpowers had ever come to nuclear war. It was not an experience that any of the participants looked forward to repeating.

Whether by skill or by circumstance, or both, the superpowers have managed to avoid war-threatening confrontations since. There have been crises, to be sure, but none of them of the magnitude of Cuba. Even the 1973 Middle East crisis, the most serious superpower clash after 1962, was by all accounts a faint echo of its famous predecessor. The participants never seem to have doubted that it would be resolved by diplomacy.[4]

But our good fortune cannot be taken for granted. As long as the East-West conflict continues, there is always the possibility that one of the superpowers, by design or miscalculation, will do something that the other is unwilling to tolerate. In recent years the probability of a war-threatening crisis may actually have increased as a result of the demise of detente and the revival of the Cold War.

No one knows how such a crisis would occur, although many of us have a favorite scenario. Mine concerns weapons in space. The United States is committed to developing a space-based defense against ballistic missiles, something the Soviets strongly oppose. Moscow has already dropped hints that it is prepared to interfere with the deployment of such a weapon system. Suppose the United States, deeply committed to the project, disregards Soviet warnings and at some point begins to put important components of a missile defense system into space. To show their displeasure, the Soviets orbit space mines in the vicinity. The United States, in turn, sends up a shuttle mission to remove or disarm the threat. But the mines, having been salvage fused (set, that is, to go off if tampered with), explode and kill six astronauts. . . .

The point about this or any other scenario is not its inherent plausibility but the fact that such hypothetical situations are taken seriously by the U.S. foreign policy and defense establishments. Such possible futures form the basis of political-military simulations used for planning and training purposes. The prevalence of these simulations implicitly recognizes that a serious superpower confrontation remains a real if disturbing possibility. So it is important to inquire how such a confrontation could occur, how the superpowers would be likely to respond to it, what our chances are of emerging unscathed, and what, if anything, can be done to enhance them. These are the questions this book will address.

The Cuban missile crisis is an obvious starting point for an analysis of this kind. It is our best example of an acute superpower confrontation, and I shall make frequent reference to it. However, Cuba is also misleading in important ways. Kennedy's success in avoiding war *and* getting the Russians to back down has encouraged the belief that this is a feasible objective in future confrontations. It has also led responsible political analysts to exaggerate the efficacy of military capability and demonstrations of resolve. In the decade following Cuba, American scholars legitimized both expectations. Political scientists enshrined the need for resolve at the core of widely accepted theories of deterrence and compellence, while decision-making theorists idealized Kennedy's handling of the crisis and held it out as proof that confrontations of this kind could be managed ably and successfully.[5]

Even without these questions of interpretation there would be important structural differences between Cuba and any future superpower crisis of the same magnitude. These differences would make such a crisis very much more difficult to resolve. In 1962 the United States had overwhelming nuclear superiority and conventional superiority in the Caribbean, the arena of the confrontation. Cuba was all one-sided in terms of military escalation. The United States brought its nuclear and conventional forces up to a high level of readiness; the Strategic Air Command was put on Defense Condition II, the only time any component of U.S. forces has ever been ordered to go to this level of readiness.[6] The Soviet Union, by contrast, refrained from any buildup, perhaps because of its disadvantage and the fear that a strategic alert on its part could trigger American preemption. But a future crisis will take place in an environment of strategic parity. The Soviets may be the ones to have the conventional advantage next time, depending on where the crisis occurs. If they do, they might not feel as constrained as in Cuba. A Soviet alert, conventional or nuclear, matching or matched by an American alert, would make any crisis more acute and correspondingly difficult to resolve. It would also make it more difficult to control.

Cuba took place in a much less sophisticated institutional environment. Many communication links were entirely ad hoc, the number of players was relatively small, and their roles were often undefined. Strategic weapons, mostly bombers, were slow and recallable. In the years since, the command and control systems of both superpowers have grown enormously in size and complexity. They have also become tightly coupled. Strategic weapons put more stress on these systems. They are capable of striking the adversary's political leadership and strategic forces in a matter of minutes instead of hours. They are also less tolerant of error, because missiles, unlike bombers, cannot be recalled. All of these developments mean less tolerance for the kinds of incidents and mishaps that plagued Kennedy's handling of Cuba. But there is reason to believe that the organizational complexity of the 1980s nevertheless makes mishaps more rather than less likely.

Today's strategic environment requires a significant shift in our ways of thinking about international crises. Leaders must show more profile and less courage. They must be less concerned with "winning" and more concerned with controlling crises, because the principal danger is no longer that the adversary will get his way but that one or both of the protagonists will set in motion a chain of events that will lead to an undesired and catastrophic war. But a conceptual shift of this kind has not occurred. Political leaders and their advisers still give every indication of believing that crisis management consists of controllable and reversible steps up a ladder of escalation, steps taken to moderate an adversary's behavior. Even as well-known a "dove" as Edmund Muskie, who played the president in a nationally televised crisis game in 1983, demonstrated his willingness to threaten the Soviet Union with nuclear weapons and, presumably, to carry through on the threat if necessary. This is precisely the kind of behavior which, whether by accident or design, increases the likelihood of war.

This dangerous political orientation is complemented by an equally fruitless and counterproductive technical approach to the problems of crisis management. Crisis management in the United States bears a disturbing resemblance to the ancient art of alchemy. Alchemists of old sought to transmute base elements into gold by simple chemistry and magical incantation. They failed because their quest was based on a false premise; elements cannot be transmuted by chemical processes. The transmutation of metals requires alteration of their nuclear structure, something that has become technically feasible only in the last fifty years. Even today, it requires vast amounts of money and energy to produce even the tiniest amount of a heavy element.

Government officials, and many academic researchers, have embarked upon a similarly fruitless quest for the secret keys to nuclear crisis management. Convinced, as were the alchemists before them, that their goal is attainable, they search for the modern day equivalent of the philosopher's stone: the organizational structures and decision-making techniques that will transmute the dark specter of nuclear destruction into the glitter of national security. Once again, the task is hopeless. Good crisis management cannot be fabricated from communication nodes, computer software, and special action groups. It requires fundamental changes in the force structures, the doctrines, and the target sets that define contemporary nuclear strategy. Like transmutation, crisis stability is theoretically possible, but for the foreseeable future it lies beyond the power of political alchemists.

Nuclear crisis management may be an oxymoron, but this does

not mean that the behavior of leaders is irrelevant to what happens in crises—only that a narrow focus on the techniques and technology of crisis management is unlikely to result in significantly improved performance. Good decision making is associated with certain underlying conditions. The most important of these are a relatively open decision-making environment, a consensus within the policy-making elite with regard to fundamental political values and institutional procedures, and freedom from the kinds of domestic political pressures which compel leaders to pursue risky foreign policies. Good decision making in crisis also demands that leaders have a working knowledge of the details of military planning and operations. None of these conditions can be created by fiat while a crisis is unfolding. If they exist at all, they are the product of extensive precrisis efforts that leaders initiate to educate themselves and to build an environment conducive to vigilant information processing and responsive policy implementation. All too often, success in this regard depends on fortuitous historical and political circumstances over which leaders have relatively little control.[7]

A shift is also required, therefore, in the focus of contemporary research and planning for crisis management. Too often, such efforts are directed at improving only the external environment in which policy makers operate, by providing greater or more rapid access to key information and individuals. Even those efforts which focus on the actual process of decision making and implementation tend to limit their horizons to the course of the crisis itself. By doing so, however, they look only at the tip of the proverbial iceberg. How a political system performs in crisis is a function of personal, group, institutional, and cultural patterns and interactions that were established long before the onset of the crisis. The most pressing task for those who study crisis management, therefore, is to delineate more clearly the precise nature of the links between performance and these underlying factors. Only then can really productive efforts be made to improve the quality of decision making.

We must also recognize that good decision making is not the same thing as successful decision making. The two are related but not synonymous, a host of other conditions also affect crisis outcomes. Among the most important are the degree of concern of leaders on both sides to avoid war, the latitude leaders have to explore a wide range of possible settlements, the time pressures they confront, the nature, gravity and compatibility of the issues at stake, and the level of military escalation and the threat of war associated with it. Any one of these conditions can compel leaders to settle for a less than favorable outcome or prevent resolution of the crisis altogether—whatever officials on both sides might desire.

Stress, organizational complexity, and time pressure are likely to characterize any future superpower confrontation. They will certainly make good decision making essential—but at the same time, it will be less of a determinant of the outcome. The reason for this apparent contradiction lies in the particular strategic, political, and psychological factors that would shape superpower interactions. Those factors will make it very difficult for leaders to resolve such a confrontation *regardless* of their skill and mutual commitments to avoid war. This book is devoted to demonstrating why this is so. It identifies the most important pressures and constraints that will act upon superpower leaders in an acute crisis and explores the ways in which they could be expected to hamper its resolution.

This is not the first book to expound the dangers of superpower crises and some of the structural reasons for it. Most of these studies have emphasized the critical vulnerability of strategic command and control and the unsettling dilemma this creates for leaders.[8] This critical aspect of nuclear strategy has been until quite recently neglected. In 1981, Desmond Ball wrote the first work to describe the architecture of the U.S. strategic command, control, and communication (C^3) system and to demonstrate its extreme vulnerability to sabotage and nuclear attack. Since then Bruce Blair has documented the vulnerability of C^3 to Soviet attack from the early 1960s to the present day. Both authors wanted to influence strategic policy. By demonstrating the vulnerability of American C^3 to nuclear assault, Ball hoped to demonstrate the absurdity of controlled, protracted nuclear war, a capability deemed essential by current U.S. strategic doctrine. Blair had an even more far-reaching objective in mind: he sought to shift American attention away from its decades-old fixation on the possible vulnerability of its strategic forces to the greater and more significant vulnerability of the command and control of those forces. He thereby hoped to win support for extensive measures to enhance the survivability of C^3.

John Steinbruner and Paul Bracken have been more concerned

with the implications of C^3 vulnerability for war-prevention in crisis. In an article published in 1981, Steinbruner drew attention to one of the policy dilemmas created by C^3 vulnerability, and in particular by the vulnerability of the political leadership to destruction at the outset of hostilities. If policy makers feared they might not be able to retaliate to nuclear attack, he argued, they would have a strong incentive to shoot first in circumstances where nuclear war appeared likely. Bracken highlighted another destabilizing aspect of contemporary command and control: the difficulty of controlling nuclear forces once alerted, and the risk of accidental war that such alerts accordingly entail. In his 1983 study he described the evolution and growing complexity of U.S. strategic C^3. Drawing upon insights from organization theory, Bracken warned that the American-Soviet strategic warning and response systems had a propensity to perform in crisis situations in unanticipated and possibly catastrophic ways.

These several works have based their analyses on an examination of the physical machinery and organizational structure of American command and control. They go on to describe strategic imperatives these give rise to and the ways in which such imperatives could hinder resolution of a crisis. Studies of this kind constitute an important first step toward understanding the causes of crisis instability. It would be wrong, however, to suppose that strategic problems can be entirely, or even best, understood in terms of engineering or organizational criteria. There is more to command and control than its technical aspects and more to crisis management than command and control. Like all strategic issues, crisis management has critically important political and psychological dimensions. To date, these remain largely unexplored.

Nuclear Crisis Management analyzes some of the political and psychological components of crisis in the expectation of offering new insight into conflict management. The findings of others which I have already mentioned are the starting point of my analysis, but I also exploit the research of scholars in other fields whose work, in my opinion, is particularly germane to the problem of crisis instability.

At least some of the "strategic realities" responsible for crisis stability are, my analysis demonstrates, as much political and conceptual as they are structural. The prince aptly opines in *Hamlet* that "There is nothing either good or bad, but thinking makes it so." This is to a marked degree true of nuclear strategy. Strategic force vulnerability is an obvious case in point. Minuteman vulnerability has not decreased one iota in recent years; presumably it has become worse as Soviet missiles have improved their accuracy. But the problem has receded in importance, to a large degree because of the successful effort of the Scowcroft Commission in 1983 to disparage its strategic significance.

An example more specific to the realm of crisis concerns the attractiveness of preemption in a confrontation in which war appears imminent. A thorough examination of the pros and cons of preemption, undertaken in Chapter 2, reveals that there are no circumstances in which preemption is an advisable policy. Both the fear that retaliation would be impossible and the advantages that preemption, some allege, confers can be shown to be greatly exaggerated. Conversely, faulty and incomplete understanding of the problems of loss of control and miscalculated escalation has led strategists in and out of government to underestimate the threats these problems would pose to crisis resolution. Chapters 3 and 4 document how and why this is so.

Although pessimistic in its assumptions, this book is not fatalistic in its conclusions. There are things that can be done to reduce crisis instability, and with it the likelihood of war. Better understanding of the dynamics of crisis could reduce the attractiveness of preemption and of the kinds of policies that court loss of control or miscalculated escalation. It could also sensitize policy makers to the need to prepare themselves and their subordinates to cope with the possibly debilitating stress that superpower nuclear crisis would generate. Although I shall make a number of technical and political recommendations, my primary goal is to encourage a conceptual shift in thinking about the nature of crisis and the appropriate means of coping with it.

The policy recommendations that grow out of my analysis are not offered as any kind of panacea. There are important structural causes of crisis instability that would be not the least bit affected if both superpowers were to adopt more enlightened approaches to crisis. The threat to peace posed by crisis instability can be overcome only by far-reaching changes in force structure, strategic doctrine, and targeting policy—the structural factors primarily responsible for the problem. Unfortunately, the current political climate does not appear conducive to the kinds of measures, unilateral or negotiated, that crisis stability would require. All the more reason, then—at least for the time being—to focus attention on conceptual aspects of the problem which may be more amenable to influence.

The Causes of War

No serious student of strategy believes that World War III will start as a gratuitous act of nuclear aggression.[9] Only in paperback adventures does the leader of one of the superpowers wake up one morning, decide that the correlation of forces is highly favorable, and give the order to push the button.[10] The incredible destruction that nuclear war would bring about, even to the side that struck first, is far too great to warrant its consideration as a rational act of policy. Leaders on both sides know this. So do their national security advisers and military establishments. They are all in accord that nuclear war would constitute the greatest disaster in the history of mankind.

Nor is a superpower nuclear war likely to come about as the result of an accident, an act of terrorism, or the efforts of a third party to start one. For reasons that I discuss in Chapter 3, the prospect of an accidental launch when forces are on day-to-day alert is exceedingly remote. Nuclear terrorism is an even more unrealistic scenario. Despite the appeal the subject has for popular fiction writers, nuclear weapons and delivery systems are not easy objects to steal. Terrorists who overcame this basic hurdle would still have to find some way of arming the weapons in the absence of the necessary code. Even if they managed to do so, it is unclear how an act of nuclear terrorism would prompt one superpower to attack the other.

Nuclear war orchestrated by a third party also seems improbable, although the idea has from time to time aroused the concern of the superpowers.[11] The usual scenario involves a nuclear attack on one superpower by another power that attempts to make the attack appear the work of the second superpower. The deception would be difficult to achieve, however, because the United States and the Soviet Union can track incoming reentry vehicles (RVs) and bombers and determine where they came from. A better possibility would be a submarine launched missile, but even that would be likely to have a telltale MIRV separation pattern. Moreover, a limited nuclear attack, regardless of what form it took, would be certain to arouse suspicions because it is the very last thing that either superpower expects the other to do.

If there is a World War III, it is most likely to be the result of either a miscalculation or an act of desperation in a crisis or conventional war. The paradigm of such a situation is 1914. World War I was brought about more by accident than by design; it was the largely unanticipated outcome of a series of amplifying miscalculations by Austria-Hungary, Germany, Russia, France, and Britain. It might have been prevented by a greater realization on the part of European leaders of the consequences of mobilization. At the same time, we would be wrong to suppose that the war was entirely attributable to poor leadership; more astute statesmen would also have found it difficult to control the cascading events of July 1914. The domestic political situation of several of the powers generated strong pressures in the direction of war while their elaborate military plans imposed severe constraints on the power and imagination of leaders.

Although there are important differences between 1914 and today, the origins of World War I hold significant lessons for hawks and doves alike. For hawks, the Great War offers insight into how and why an adversary can be tempted to exploit his perceived political-military advantages. The crisis began as an attempt by Austria-Hungary, backed by Germany, to use the assassination of Archduke Franz Ferdinand as a pretext for destroying Serbia. German and Austrian leaders expected that Russia, Serbia's principal backer, would stand aside rather than confront their combined military might. If Russia nevertheless entered the fray, they were convinced that France, Russia's ally, would remain neutral. The July crisis was therefore envisaged as the catalyst for a localized Balkan war.

In point of fact, Germany and Austria made a grievous miscalculation. Not only did Russia back Serbia to the hilt, but France proved steadfast in its support of Russia. Because of this, Austria's démarche triggered a European war. Germany, much to its horror, had to fight Britain as well as Russia and France, because of its ill-advised invasion of Belgium. For doves, 1914 accordingly sheds light on how attempts at coercion can have unforeseen and disastrous consequences.

Better understanding of the dynamics behind both brinkmanship and miscalculation would help us manage adversarial relationships and the conflicts they spawn in a more sophisticated and successful manner. One way to do this would be to explore the role of both phenomena in 1914 in an attempt to mine that case for more widely applicable lessons of conflict management. As I have previously done with regard to brinkmanship, I focus here on miscalculation and its causes. In succeeding chapters I analyze three causal sequences to war, all of which were instrumental in bringing about war in 1914. Those chapters describe the underlying or structural conditions associated with each of these sequences in 1914. They seek to ascertain the extent to which those conditions exist today and could be expected to shape the course of some future superpower confrontation.

The first of these sequences is *preemption.* It occurs when one of the protagonists in a crisis goes to war because its leaders believe that their adversary is on the verge of doing so. When hostilities appear unavoidable, preemption can be an attractive option if it is believed that striking first will confer a significant or even decisive advantage. It is entirely conceivable that mutual recognition of this asymmetry could prompt one or other of the protagonists to go to war even though both would prefer to keep the peace. Thomas Schelling has called this "the reciprocal fear of surprise attack." He reasons:

> If surprise carries an advantage, it is worthwhile to avert it by striking first. Fear that the other may be about to strike in the mistaken belief that we are about to strike gives us a motive for striking, and so justifies the other's motive. . . . It looks as though a modest temptation on each side to sneak in a first blow—a temptation too small by itself to motivate an attack—might become compounded through a process of interacting expectations, with additional motive for attack being produced by successive cycles of "He thinks we think he thinks we think . . . he thinks we think he'll attack; so he thinks we shall; so he will; so we must."[12]

Until quite recently, most of the literature conceived of crisis stability in terms essentially similar to Schelling's.[13] Analyses of this kind for the most part ignored the possibility that war could break out in the absence of any deliberate decisions by leaders on either side to fight it. They failed to consider our second sequence to war, *loss of control.*

Loss of control can take a variety of forms and can have diverse causes. It can result from fragmented political authority, domestic pressures that leaders are powerless to resist, or an institutional malfunction or breakdown. It can also be the inadvertent and unanticipated outcome of military preparations taken to protect oneself in a crisis or to convey resolve to an adversary. As 1914 revealed, interacting organizational routines can lead to war in a quasi-mechanical manner. Luigi Albertini and A. J. P. Taylor, two of the most eminent historians of that conflict, considered this to have been one of its most important proximate causes. Political leaders, they argued, in effect "made" decisions for war without realizing at the time that they were doing so.[14]

Contemporary organizational theorists argue that World War III could start in the same way. Paul Bracken warns that an amplifying feedback loop could develop between the warning and response systems of the superpowers as a result of any significant military preparations by either one of them. Suppose the United States went on strategic alert in order to demonstrate resolve in a crisis; "the Russians would monitor some of the changes taking place in American monitoring operations and in reaction might speed their preparations for attack. America might respond in kind. What started as mere brinkmanship," Bracken fears, "might end in nuclear war."[15]

A third sequence to war is *miscalculated escalation,* when one of the adversaries crosses the other's threshold to war in the false expectation that his action will be tolerated. The Russian mobilization is a classic example. In the opinion of most historians, it was the step that made war all but unavoidable. However, Russian political leaders mobilized in 1914 in the belief that mobilization would be a deterrent to war. The American decision to invade and occupy North Korea in 1950, Nasser's remilitarization of the Sinai in 1967, and Argentina's occupation of the Falklands in 1982 have also been described as instances of miscalculated escalation.[16] In each of these confrontations, the initiator took action unlikely, he thought, to provoke a violent response. All three led to war. Little of a theoretical nature has been written about the phenomenon of miscalculated escalation, even though it is probably the most important cause of inadvertent war.

These three sequences to war are conceptually distinct. Any of them can be described separately but to demonstrate their practical

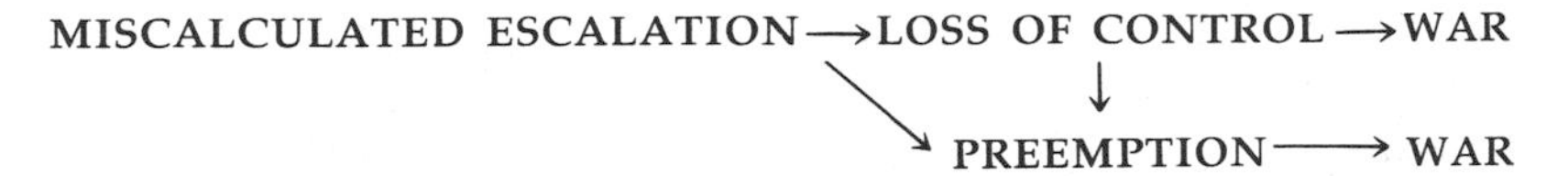

Figure 1. Possible paths to war in crisis

effect, they must be analyzed in conjunction with one another. The relevant analogy here is to epidemics, which also have multiple and interrelated causes. Among the most important of these are the presence of a pathogen, a mechanism for its spread among humans, and low human resistance because of the absence of antibodies or the debilitated state of the population. Sanitary conditions and the nature and rapidity of any institutional response will also influence the progress of the disease. It is the totality of these conditions which determine whether the disease becomes an epidemic.

So it is with crisis; the collective interaction of the three causal sequences, each with its own individual probability, determines whether war breaks out (see Figure 1). In 1914, for example, Russia's miscalculated escalation prompted Germany's decision to preempt. Loss of control, in turn, was a cause contributing both to Russia's miscalculated escalation and to Germany's preemption.

Efforts to minimize the risk of war in crisis are also interrelated. Preemption, loss of control, and miscalculated escalation have some distinctive but also some common causes. Civilian ignorance of war plans, to cite one example, can be shown to contribute to all three. Successful efforts to educate political leaders about the process of escalation and the details of military alerts would constitute an important step toward reducing the likelihood that some future superpower crisis will erupt into war. There are other measures that would have a similar ameliorating effect across the board.

Unfortunately, some of the most important ways of preventing particular sequences to war would also make other sequences more likely. This is true of many measures designed to cope with preemption or loss of control. We can reduce pressure to preempt in a crisis, for example, by reliance on a launch-on-warning (LOW) capability, predelegation of launch authority, or prepackaged targeting options to be executed automatically in response to a breakdown in the chain of command. But all three policies court loss of control; they increase the likelihood that war will start in circumstances in which a president or premier remains committed to keeping the peace.

Many measures designed to reduce the chances of loss of control entail similar trade-offs. The installation of Permissive Action Links—physical locks to prevent weapons aboard bombers and missiles from being armed in the absence of an Emergency Action Message from the National Command Authority—and the probable decision of any president to remain in Washington throughout a crisis, all make retaliation more difficult to carry out. Uncertainty about the nation's ability to respond to nuclear attack in turn generates pressure to preempt once war appears unavoidable.

Clearly, any comprehensive attempt to build greater crisis stability involves difficult trade-offs. Before decisions can be made, however, we must know more about the causes and mechanisms of each sequence to war, the relative danger that each would pose in a future crisis, and the likely impact upon all three sequences of any measure designed to reduce the chances that any one of them will lead to war.

Three Sequences to War

[2]

Preemption

This chapter and the next analyze the strategic alert and response system of the United States. The system consists of all the sensors, communication links, analysis centers, command posts, and weapons that would detect an attack against the United States and respond to it in a timely and effective manner. The military officers who direct this system have two primary concerns. The first is the need to maintain the capability to retaliate against any Soviet attack. This is deemed essential not for any punitive reason but as the surest means of deterring such an attack in the first place. Deterrence is, of course, the justification of all American strategic forces. The second concern is the need to prevent accidental or unauthorized use of nuclear weapons. Public officials have long recognized that the incredible destructiveness of such weapons demands extraordinary measures to ensure their safe storage and handling and to guarantee that only the president, or other properly constituted authority, can initiate their use. As I shall show, these concerns have shaped the complex physical and institutional structure of the U.S. alert and response system and the principles that govern its operation.

My analysis reveals that despite elaborate preparations and precautions, military officials remain uncertain about their ability to retaliate against a Soviet attack. By contrast, they seem confident, if not smug, about their ability to prevent an accident or unauthorized launch. The policy implications of these judgments are profoundly disturbing.

Doubts about the feasibility of retaliation encourage officials to consider more extreme measures to enhance its likelihood. In particular, they raise the prospect in an acute crisis of preemption—that is, a first strike initiated when war appears unavoidable. Preemption, according to some strategic pundits, could confer two additional advantages: it could limit damage to the attacker's homeland and gain him a significant postwar military advantage. In this chapter I take issue with these arguments. Superpower nuclear preemption would be an inappropriate and entirely disastrous course of action in any circumstances.

Many of the measures currently relied on to facilitate retaliation put a strain on the procedures and safeguards erected to prevent accidental, unauthorized, or inapproprite use of nuclear weapons. The next chapter documents this assertion and spells out some of the ways in which war could start in a superpower crisis despite the efforts of both sets of leaders to prevent it. The officers in command of the strategic alert and response system are generally unwilling to recognize this danger. In part, this unwillingness stems from their fixation on retaliation and their corresponding reluctance to recognize the risks inherent in measures designed to guarantee the ability to retaliate.

My overall conclusion is that strategic reality is very different from the way the strategic establishment perceives it. To be sure, retaliation is not 100 percent assured, but American officials and civilian strategic analysts have grossly exaggerated the negative political-military consequences of this uncertainty. The more serious danger is war arising through loss of control. The risk, moreover, is likely to be made much more acute than it need be by the kinds of measures which would be taken in a war-threatening crisis to ensure retaliation. We could therefore take an important step toward war prevention by reversing the priorities of our strategic policy.

Fear of Strategic Disadvantage: 1914 and Today

I start with an examination of the role preemption played in the outbreak of World War I. All the incentives for preemption in 1914, I shall show, rested on erroneous assumptions about the nature of a future European war. History threatens to repeat itself: the political-military elites of the United States and the Soviet Union, I argue, appear to share many of the false expectations that were responsible for German preemption in 1914. They too are committed to offensive strategies; they too give evidence of believing in the importance of striking the first blow. Expectations of this kind constitute a grave threat to peace today just as they did on the eve of World War I.

Military and political leaders in 1914 assumed, with few exceptions, that any European war would be short; general staffs prepared on average for a conflict that would last six weeks. They also believed that the initial clash of opposing armies, something meticulously planned for in the elaborate mobilization schemes of all of the powers, would decide the outcome. For this reason, generals throughout Europe prepared to mobilize as quickly as possible in response to news of similar preparations by adversaries.[1]

Fear of the consequences of caution and restraint was greatest in Germany because of the nature of its war plan. The general staff had concluded some years earlier that Germany would have to fight France and Russia in any European war. As they doubted Germany's ability to wage war simultaneously on two fronts, they looked for some way of dealing with their enemies sequentially. According to the war plan General Alfred von Schlieffen initially devised, France, the stronger of the two adversaries, was to be attacked first. A delaying action would be fought in the east against the more slowly mobilizing Russians, but the full weight of German military might would turn against Russia after France had capitulated.

Schlieffen's plan staked everything on Germany's ability to knock France out of the war quickly; speed was essential to free German armies for redeployment in the east in time to halt the Russian steamroller before it leveled Berlin. Schlieffen devised a daring strategy: almost all of the standing army and mobilized reserves were to be thrown into the battle against France, leaving only a thin screen of covering forces in the east. Schlieffen made an equally fateful decision because the Franco-German frontier was largely unsuited to the rapid advance of large armies: he would direct the main axis of the offensive through Belgium. The German army was to overrun that country and then wheel south in order to outflank and encircle the French Army behind Paris.[2]

Prewar German calculations called for the defeat of France and the redeployment of the army to the east by the fortieth day of mobilization. The timetable left hardly any room for slippage, because the general staff estimated that not long afterward the Russian Army, meeting only token opposition, would fight its way to Berlin. The need to win a quick victory over France was one of the considerations uppermost in the minds of German generals throughout the July crisis. They insisted that neither France nor Russia should be permitted to hinder the success of the Schlieffen Plan by stealing a march on Germany with respect to mobilization. On 30 July, when Chief of Staff Helmuth von Moltke learned of the Russian mobilization, he was adamant that Germany mobilize immediately. At his insistence, an ultimatum was sent to Russia demanding revocation of the order to mobilize within twelve hours. When this deadline expired, Germany went to war.[3]

Germany was in effect stampeded into war by fear of the consequences of restraint. In retrospect this was a tragedy of monumental proportions, because the worst fears of the German military proved entirely unjustified. The Russian offensive, which got under way earlier than Berlin had anticipated, turned out to be much less of a threat than had been supposed. The Russian armies advancing into East Prussia were decisively defeated at the battles of Tannenberg and the Masurian Lakes by the relatively meager covering forces the general staff had left in the east.

The precipitous nature of the Russian offensive, undertaken to ease the German pressure on France, contributed to the magnitude of its failure. The offensive had carried Russian forces deep into East Prussia before they encountered any serious opposition. Russian logistics and command and control, archaic even by the standards of the day, broke down under the strain of this advance. The resulting sluggishness and poor coordination of Russian forces permitted the greatly outnumbered German defenders, who profited from internal lines of communication and a dense and well-exploited railway network, to defeat piecemeal the two prongs of the Russian offensive.[4]

Another irony of the campaign in the east was the abortive German attempt to reinforce General Paul Hindenburg with forces drawn from the west. The Russian invasion touched off a panic in Berlin, especially among the Junker class, who feared that the advancing enemy would overrun their estates in East Prussia. In response to this threat, two corps were withdrawn from the extreme right of the army group in Belgium and dispatched posthaste by rail to the east. These forces proved to be of no immediate use;

by the time they arrived, the Russians had already been defeated. But some historians have suggested that their absence from the Western Front may have been decisive, for they would have been well-situated to have turned the flank of the British Expeditionary Force.[5] Had that happened, Paris might have fallen.

Current strategic "reality" for the Soviet Union comes close to resembling the German predicament in 1914. In Europe the Soviets are committed to an offensive conventional strategy. Success depends upon exploiting a numerical advantage to win a quick victory over economically superior adversaries. The mobilization and reinforcement rates of the two alliance systems make it imperative that the Warsaw Pact launch its offensive before NATO is fully mobilized. The result is tremendous pressure on th Soviet Union to make the decision for war in response to any NATO decision to mobilize. Like their counterparts in the Germany of 1914, Soviet leaders could opt for war in an acute crisis because they fear the military consequences of allowing their adversaries to proceed unimpeded with military preparations.[6]

NATO mobilization would presumably occur only in response to unambiguous evidence of Soviet mobilization. Allied mobilization under any other circumstances is politically inconceivable; West European governments are simply too reluctant to do anything that would risk war or that their political opponents could portray as unnecessarily provocative. The Soviets would probably accompany mobilization with an extensive propaganda campaign designed to discourage a NATO countermobilization by playing upon these European fears. A mobilization that grew out of a crisis in Eastern Europe might well be accompanied by assurances, sent through public and private channels, that Moscow was planning only to invade one of its own allies. Soviet diplomats would be almost certain to warn nervous West European allies that NATO mobilization would probably make continental war inevitable.[7]

A situation of this kind would confront NATO with an awkward choice. To forgo mobilization would leave the Western alliance exposed to Soviet attack and possibly even tempt that attack. But mobilization would risk making NATO's fears of war self-fulfilling by forcing Moscow's hand—for the Soviets need to strike before a NATO buildup leads to a less favorable balance of forces on the Central Front. Fear of strategic disadvantage would almost certainly prompt NATO military authorities to push for mobilization, especially if they detected inconsistencies between the Warsaw Pact's military preparations and Moscow's declarations of intent. Would elected officials act on their request? The answer would depend very much on the political circumstances at the time.

A similarly destabilizing dynamic operates with respect to theater nuclear weapons. Current Soviet strategy appears to call for holding Soviet nuclear weapons in reserve in relatively safe rear areas while using conventional forces, and possibly chemical weapons, to destroy or disrupt NATO's theater nuclear capabilities.[8] The Soviets would probably launch a nuclear strike only if it appeared that NATO was on the verge of doing so. This strategy, known to the West, is certain to prompt greater efforts by NATO to protect its nuclear forces and their command and control. However, measures to protect those nuclear forces, by taking them out of garrison and dispersing them in the countryside, could appear to the Soviets as preparations for a Western nuclear strike.

The Soviet dilemma is made more acute by geographical asymmetries and by the capabilities of Western theater nuclear systems. Soviet spokesmen allege, not without some justification, that the Pershing IIs deployed in Western Europe will provide NATO with a potent first-strike weapon. Pershing IIs, the Soviets claim, could reach Moscow six to eight minutes after being launched from their bases in West Germany. With an estimated range of 1800 kilometers and an expected circular error of probability of about 30 meters, the P-II appears especially well-suited, if not deliberately designed, for theater wide counterforce missions. It would be particularly effective against Soviet missile systems and their command and control centers in Eastern Europe and the western military districts of the USSR. For this reason, P-II deployment is likely to push the Soviets further in the direction of adopting a "hair trigger," launch-on-warning posture. Worse still, the presence of P-IIs may tip the balance in favor of Soviet preemption in an acute crisis or a conventional European war.[9]

Fear of strategic disadvantage is also a destabilizing factor at the intercontinental level. Both superpowers, it is well known, are committed to war-fighting doctrines that place great emphasis on highly accurate land-based missiles and their command and control. These are also the most vulnerable strategic assets either superpower possesses. Command, control, communications, and intelligence, commonly referred to as C^3I, makes an extremely attractive target because of its value as a force multiplier. Proper and timely use of C^3I would permit the initiator of a strategic exchange to lay down weapons in a more complex and accurate attack pattern, one that would make them more likely to destroy their targets. It would also facilitate post-attack assessment. This, coupled with a retargeting capability, would also allow strategic forces to be used more effectively and efficiently against surviving enemy forces.[10]

C^3I is crucially important to war-fighting ability. It is also extremely vulnerable to attack. That vulnerability provides a strong incentive for striking first if war appears unavoidable or even highly likely. Bruce Blair, author of the authoritative study of U.S. strategic command and control, offers the judgment that "mutual command vulnerability creates strong incentives to initiate nuclear strikes before the opponent's threat to [one's own] C^3I could be carried out." For this reason, Blair concludes, "both sides would come under increasing pressure to mount a preemptive attack in a crisis of increasing gravity."[11] High-ranking American military officials acknowledge this unsettling reality. As General Richard Ellis, former commander-in-chief of the Strategic Air Command (SAC), confided to an academic audience in 1982, "one of the most destabilizing things that we and the other side will have to live with is the case where one side knows the other side's command and control and communications system, or weapon system, is vulnerable. That," according to the general, "is an incentive for the side living with that vulnerability to go first."[12]

U.S. Command and Control

If we are to evaluate the threat posed to crisis stability by mutual pressures to preempt, the first step we need to take is to determine the relative vulnerability of U.S. and Soviet C^3I. It is with this end in mind that I review the structure of the U.S. system. Some assessment of "connectivity," the ability of national command authorities to determine the status of and transmit orders to the various elements of their strategic forces, is also essential to the next chapter's analysis of control problems.

Early Warning

The development of intercontinental ballistic missiles (ICBMs) and submarine launched ballistic missiles (SLBMs) has dramatically increased the vulnerability of the United States to attack. ICBMs, launched from missile fields in southern Russia, would take between eighteen and twenty-two minutes to reach their targets in the continental United States. Ballistic missiles fired from offshore submarines would arrive over their targets in five to eight minutes. To cope with this threat, the United States relies upon a combination of satellites and radar installations to give prompt warning of attack. Some warning might also be received in advance of an attack. Satellites, aircraft, and ground-based listening posts would likely pick up a significant increase in communications as well as changes in their patterns and frequencies as the Soviets switched from peacetime to wartime operating modes. They might also intercept ICBM launch orders or other information concerning attack preparations.[13]

Tactical, or actual, warning of attack would be provided by a multitiered system.[14] The first element of this system, three Defense Support Program (DSP) satellites in geosynchronous orbit, are equipped with heat-sensitive infrared telescopes to detect the hot exhaust plumes of Soviet ICBMs in their boost phase. Within two minutes of any launch, these satellites would alert the North American Aerospace Defense Command (NORAD) in Cheyenne Mountain near Colorado Springs. They would also alert other important national command centers, among them the National Military Intelligence Center and the National Military Command Center in the Pentagon and the White House Situation Room.

Warning of attack would also be signaled by ground-based back scatter Over-the-Horizon radars. There is one in Maine, another being built on the West Coast, and a third planned for the southeast. With the most favorable ionospheric conditions, these facilities could detect ICBM launches at about the same time as the satellites. Under less favorable conditions, they would provide warning within five to ten minutes of launch.

Fifteen minutes after Soviet launch an attack would be confirmed by the Ballistic-Missile Early Warning System (BMEWS) radars in Alaska, Greenland, and the United Kingdom. COBRA DANE, a phased array radar in the Aleutians, would also provide warning at about this time. It would convey information about the number of the incoming reentry vehicles and their possible targets. The best attack characterization would come from the Perimeter Acquisition and Raid Characterization System (PARCS) in Grand Forks, North Dakota.[15]

Warning about offshore submarine launched missiles would be provided by Western Hemisphere DSP satellites augmented by phased array warning radars (PAVE PAWS) in Massachusetts, Florida, and California. Two additional PAVE PAWS are under construction, in Georgia and Texas. Work is also under way to upgrade the existing PAVE PAWS to detect smaller and more distant targets. To warn against bombers, the United States currently relies on the Distant Early Warning (DEW) Line of thirty-nine unattended and thirteen minimally attended radars. The DEW Line will eventually be replaced by the North Warning System. All of these systems are capable of making some determination of the number and probable targets of the incoming warheads.

Negative Control

In peacetime the primary objective of the U.S. command and control system is to maintain "negative control" over all nuclear weapons and their delivery systems, in order to prevent an accidental or unauthorized launch. The principal means of doing so is institutional. Most important alerting actions and emergency procedures can come into effect only in response to the declaration of the requisite alert level by two separate and independent commands. Institutional checks are built into every level of the system, down to the "two-man rule." Nuclear weapons in missile silos or submarines can be launched only by the coordinated action of two or more individuals at consoles spaced far enough apart (reportedly twelve feet from each other) to prevent one person from operating both at the same time.[16] A further precaution, the Permissive Action Link (PAL) program, consists of physical locks designed to prevent any air force or army nuclear weapon from being armed in the absence of a special numerical code. The National Command Authority (NCA) would transmit this code to weapons units in the form of an Emergency Action Message (EAM), also known as the go-code.[17]

Both these safeguards apply to the bomber force. In response to warning of an attack, B-52s and FB-111s on strip alert would be ordered to take off and fly to predetermined positions well outside Soviet airspace. There they would circle and await receipt of an EAM. Nuclear weapons could be armed only after this message was received and authenticated, and even then arming would require the simultaneous efforts of several crew members. On missile-launching submarines (SSBNs) there are no PALs, but many more individuals, officers and crew members, are involved in the arming and launching of nuclear missiles.[18]

Negative control entails an obvious wartime cost. Because it requires central authorization for all nuclear arming and release, it is a source of strategic vulnerability. An adversary that could prevent the go-code from reaching American strategic forces, by jamming, spoofing, or destroying critical elements of the command and control system, might thereby forestall retaliation. For this reason, the NCA has been ensured access to forty-three different communication systems across the electromagnetic spectrum, all of them part of the World Wide Military Command and Control System. These systems include radio, microwave relays, ground and underwater cables, and military and civilian telephone systems. The most survivable portions have been designated the Minimum Essential Emergency Communications Network. But even these systems are vulnerable as, of course, is the NCA itself.[19]

C^3I systems and their components are vulnerable to destruction or disruption by physical and electronic attack. Many of these systems are especially vulnerable to the direct and indirect effects of nuclear weapons. Most communications systems incorporate vital above-ground elements, such as radars, transmitting towers, switching stations, receiving antennae, and telephone poles. These "soft," fixed targets can be readily destroyed. In some instances their functioning can be disrupted by nuclear explosions thousands of miles away. Such detonations can black out long-distance radio communication for several hours, because the ionization they produce interferes with the transmission of both ultra-high-frequency and high-frequency broadcasts.[20]

Explosions high in the atmosphere also generate a short-lived electromagnetic pulse (EMP) that can burn out electric or electronic circuits, especially those that rely on computer chips. Large nuclear weapons set off at an altitude of 200 miles would put out of operation most unshielded electrical and electronic equipment within a radius of 750 miles. Several such weapons detonated at the right locations above the United States could not only degrade communications but incapacitate the national power grid for hours or even days. They would render inoperable much of the military's C^3I network.[21]

Satellites are also vulnerable to EMP and to other indirect effects of nuclear explosions. X-rays, gamma rays, and neutrons can disrupt or damage satellite circuitry, while ionizing radiation in the atmosphere can interfere with or even prevent satellite communication with ground stations.[22] Low-altitude satellites are increasingly vulnerable to antisatellite weapons, expected to become much more capable in the course of the next decade. The satellites the United States depends on for communications, navigation, and early warning are in geosynchronous or near-geosynchronous orbit, where they are much less vulnerable than satellites in lower orbits. For the time being, they are well beyond the range of conventional antisatellite weapons. In theory they could be destroyed by nuclear explosions set off in their vicinity, but their destruction could prove a difficult and time-consuming task to accomplish in wartime. It would be much more feasible to attack the small number of vulnerable ground-based control and receiving stations on which the utility of these satellites depends.[23]

C^3I Modernization

The United States is attempting to "harden" many existing communications facilities and to deploy new systems that will be less vulnerable to attack. The air force is building the Ground Wave Emergency Network (GWEN) that will use numerous relay stations to transmit low-frequency signals between SAC and its bomber and missile force. The network will use sophisticated communications technology to shunt digitized messages in discrete packets along a variety of alternative routes. To knock out this system, the Soviets would have to destroy a great percentage of its relay stations. Low-frequency also has the advantage of being very difficult to jam, and it is less vulnerable than other kinds of transmissions to nuclear effects. Unfortunately, GWEN will not be able to carry messages directly to submarines or bombers because of the limited range of its signals. The navy and air force will still depend for this purpose on shore-based very-low-frequency radio stations and other equally vulnerable transmitting facilities.[24]

In space the military is deploying a new generation of communications satellites. These DSCS-III (Defense Satellite Communications Systems) satellites will be more resistant than their predecessors to jamming, electromagnetic pulse, and transient radiation effects on electronics. By the end of the decade the space shuttle will place the Military Strategic, Tactical, and Relay (MILSTAR) constellation of satellites into orbit. These satellites are meant to be capable of functioning throughout a nuclear war. They will be equipped with an array of sensors to warn them of attack, and small rocket motors will permit them to take evasive action.[25]

Plans are also afoot to rectify the dependence of satellite-borne sensors on three stationary and vulnerable ground stations. Six mobile ground terminals (MGTs) are being built as backups for DSCS. Other MGTs will provide backups for MILSTAR and the Defense Meteorological Satellite Program. These MGTs will be indistinguishable in appearance from large commercial trucks, but satellite terminals, computers, transmitters, power supplies, and related equipment will be packed inside. Soviet photo-reconnaissance satellites are expected to have difficulty in locating the MGTs in peacetime. To perform their wartime mission, however, MGTs must stop and set up their equipment. Any Soviet electronic intelligence satellites still functioning could then determine their location.[26]

Continuing Vulnerability

C^3I is best envisaged as a complex nervous system. Satellites, receiving stations, radio towers, ground lines, microwave relays, and all the other hardware are the neurons that collect information about the strategic environment and the nerves that transmit messages throughout the system. In the human body some important actions can be initiated by the nervous system itself. So too can command decisions, even critical ones, be made by some of the C^3I system's far-flung ganglia. But like the body, C^3I requires a brain to give the system overall direction by "making sense" of all the information received, deciding upon appropriate responses when action is required, and coordinating the behavior of its many parts toward these ends.

The brain of the U.S. C^3I system, the National Command Authority, consists of the president, the secretary of defense, and the joint chiefs of staff. Their command posts are known as the National Military Command System and include the White House, the National Military Command Center in the Pentagon, its underground clone at Ft. Ritchie, Maryland, and the National Emergency Airborne Command Post (NEACP, pronounced "kneecap"). These headquarters are linked to one another and to numerous other command posts in the United States and overseas by redundant means of communication. NORAD, near Colorado Springs, the Strategic Air Command headquarters at Offut Air Force Base, outside Omaha, and SACLANT, in Norfolk, are among the many ganglia that exercise more direct control over discrete elements of the information-gathering system or the strategic forces themselves.[27]

All of these command centers are vulnerable. Even the hardened ones, such as NORAD, buried deep within Cheyenne Mountain, could not survive a direct hit from a multimegaton weapon.[28] In order to ensure their ability to transmit an EAM, the services have deployed airborne command posts. These EC-135s and E-4s, military versions of Boeing 707s and 747s, contain sophisticated computers and communication equipment. The 747s are being hardened against EMP. One Looking Glass, as SAC calls its secondary command post, is always airborne over the Midwest with a SAC brigadier or major general aboard. Looking Glass aircraft can launch any or all of the one thousand Minuteman ICBMs. They can also relay a launch order through launch-control aircraft that would be sent aloft in a crisis and positioned over the three principal U.S. missile fields.[29]

Looking Glass can also communicate with the bomber force by means of communications relay planes equipped with ultra-high-frequency and super-high-frequency transmitters. Such wavelengths are affected only by line of sight nuclear bursts. As a last resort, Looking Glass can launch the Emergency Rocket Communications System, twelve Minutemen with ultra-high-frequency transmitters in their nosecones, each of which could transmit the go-code to the ICBM and bomber forces.[30]

The navy has attempted to develop similar redundancy to guarantee communications with its far-flung Fleet Ballistic Missile submarine force. SSBNs, while the most suvivable leg of the strategic triad, are also the most difficult to communicate with. Day-to-day contact is maintained by satellite, very-low, and very-high-frequency radio transmissions. Very low frequency is especially difficult to jam but has the drawback of an exceedingly low rate of data transmission. Very high frequency, while more rapid, cannot penetrate more than a few meters of water, making it necessary for SSBNs to place their receiving antennae near the surface—which makes them vulnerable to detection. To minimize SSBN vulnerability, the navy is installing an extremely low-frequency transmitting station in Michigan to supplement one already operating in Wisconsin. Protected against EMP, these transmitters will permit communications with submarines that are submerged and traveling at high speed. Of course, neither transmitter could be expected to survive a nuclear attack. For this reason, the navy is experimenting with blue-green lasers as a means of beaming a very narrow signal via satellite to a submerged submarine trailing an underwater antenna.[31]

The difficulty with all of these solutions is that they rely upon ground-based components, especially transmitters, that can easily be destroyed by nuclear attack at the very outset of a war. As a backup system, the navy, like the air force, relies on airborne communications. Two squadrons of specially modified C-130 aircraft, known as TACAMOs ("Take Charge And Move Out"), are equipped with very-low-frequency transmitters. At least two of these planes are constantly aloft, one each over the Pacific and Atlantic oceans. Plans call for modified and more capable Boeing 707s to replace the TACAMOs and to provide the means to communicate with submarines via satellite. For the foreseeable future, however, these aircraft will continue to depend on fourteen ground stations to communicate with the National Command Authority. The location of these stations is undoubtedly known to the Soviets. Aircraft, their airbases, and above all the NCA will remain vulnerable to attack.[32]

Fully half of the four hundred primary and secondary C^{3}I targets in the United States could be struck within five to eight minutes by missiles fired from offshore Soviet submarines on routine patrol. C^{3}I could be disrupted even sooner by EMP produced by SLBMs detonated at high altitude during the upward portion of their trajectories. This vulnerability has led most authorities to conclude that connectivity would not last long in a nuclear war. Optimistic estimates speak of hours. But many experts believe that command

and control would break down in the opening minutes of any major attack on the United States. If breakdown did happen, however, it would not necessarily preclude retaliation; in theory an attack could be detected and verified and an EAM sent to the strategic forces before any Soviet missile arrived.

The only kind of attack that could conceivably prevent retaliation would be one aimed directly at the president and the national military leadership. Success would "decapitate" U.S. strategic forces. Admiral Gerald Miller, former deputy director of the Joint Strategic Target Planning Staff, testified in March 1976 that the United States "might have considerable difficulty" in executing a retaliatory strike if the president were killed. More recently, nongovernmental analysts, among them Desmond Ball, John Steinbruner, and Bruce Blair, have also stressed the danger of this attack scenario.[33]

In the United States the final authority with respect to nuclear weapons is the president. As commander-in-chief of the armed forces, he is at the apex of the national command authority. The president is always accompanied by a military aide carrying the "football," a black briefcase that contains the codes necessary to order a nuclear strike. To use the codes, however, the president must secure the compliance of two other high officials. The Reorganization Act of 1958 stipulates that he must give the command to use nuclear weapons to the secretary of defense, who in turn must tell the chairman of the Joint Chiefs of Staff to execute the war plan.

Because of the critical importance of the president, plans have been made to evacuate him and other high officials from Washington if a nuclear attack seems imminent. Helicopters would probably be used for this purpose. Communications with the strategic forces would be maintained via mobile command posts or a NEACP aircraft, one of which is always on alert at Grissom Air Force Base in Peru, Indiana. These Boeing 707s and 747s are also capable of accommodating the president, some of his advisers, and top military officials, providing them with an airborne refuge from which to conduct strategic operations.[34]

So much for the theory. In practice the twenty minutes allowed for presidential evacuation is totally insufficient. In a surprise test of the system authorized by President Jimmy Carter in 1977, it took forty-five minutes for the first helicopter to reach the White

House.[35] It was then almost shot down by security-conscious secret service agents. Even if the evacuation scheme could be perfected, however, it would still be unlikely to save the president. The White House, the Pentagon, and Andrews Air Force Base in surburban Maryland could all be destroyed with fewer than five minutes' warning by missiles fired from an offshore Soviet submarine.

If the president is killed, authority passes down a sixteen-person line of succession stipulated by the Constitution and American law. The line begins with the vice-president and proceeds through the speaker of the House of Representatives, the speaker *pro tem* of the Senate, and the members of the cabinet. But in all likelihood most if not all of these officials would be dead or unreachable in the aftermath of a major nuclear attack. It would be very difficult to determine who was next in the line of succession. Chaos would probably ensue, making any policy decision that required the concerted action of political authorities utterly impossible to execute. The 1981 assassination attempt on Ronald Reagan was sufficient disruption to trigger a power struggle between "I'm in charge" Alexander Haig and Vice-President George Bush. Though hardly analogous to a nuclear war, this incident nevertheless gives a hint of what the postattack political environment might be like. For this reason, analysts of C³I believe that in crisis, authority to fire nuclear weapons in retaliation would be predelegated by the president at least part of the way down the military chain of command. Although such a policy might be a successful hedge against the effects of decapitation, it can also give rise to serious problems of its own, as the next chapter makes apparent.

The only way to save the president would be to evacuate him along with other important officials, well in advance of a Soviet attack. The president might nevertheless choose to remain in Washington for compelling domestic political reasons. As an evacuation could hardly be carried out secretly, moreover, it carries with it the danger of arousing Soviet suspicions. Is the president fleeing Washington because the United States is preparing to launch a first strike? Because of what Robert Jervis has called the "masking effect"—the fact that outside observers cannot distinguish whether most military preparations are offensive or defensive in nature—such actions could prompt the Soviets to launch an attack of their own for fear of being preempted.[36] To prevent this kind of catastrophic misunderstanding, and to forestall the kind of domestic panic that an evacuation could trigger off, the president would probably decide to remain, visibly vulnerable, in Washington.[37]

Incentives for Preemption

It is important that we recognize both hope and fear contribute incentives for preemption. Hope is not something we usually think of in connection with nuclear war, but it could nevertheless play an important and dangerous role in crisis decision making. The only way to win a nuclear war is to deprive the other side of its capability to retaliate, remote as the chances of doing so may be. This can be accomplished only by striking first. Even if preemption fails to deprive the adversary of all retaliatory capability, the conventional wisdom holds that striking first could still succeed in significantly limiting damage to the initiator.

The strategic fears that encourage preemption have the same basis as the strategic hopes. The most important is the widely shared concern that retaliation might be marginal or altogether impossible in the aftermath of a massive attack that one's adversary aimed specifically at destroying personnel and facilities critical to command and control. The concern for retaliation, it should again be stressed, derives from a desire not to punish the adversary but rather to deter him. It is based on the belief that the probability of an adversary choosing preemption diminishes as one increases one's own ability to respond in kind.

American strategic doctrine stipulates onerous military requirements for deterrence. It asserts that retaliation, even massive retaliation, is insufficient for deterrence. The United States must also retain the capability to carry out coordinated counterforce strikes because Soviet leaders, it is alleged, would base their pre-attack calculation of cost and gain on the expected relative postwar military standing of the two superpowers. This assumption generates additional pressure on the United States to preempt in a crisis because such counterforce attacks would be even more difficult to carry out than massive retaliation in the aftermath of a Soviet attack. American officials fear that, because the Soviets know this, *they* have a great incentive to preempt.[38] And so it goes. . . .

These hopes and fears alike have been exaggerated by contemporary strategic analysts. Much of the conventional wisdom about preemption is either wrong, as in the case of damage limitation, or based on erroneous or inappropriate conceptions. These conceptions, as much as or possibly even more than objective strategic realities, are the source of pressure to preempt in a crisis.

Guaranteeing Retaliation

Doubts about the feasibility of retaliation derive from recognition of just how vulnerable leadership cadres and communication nodes are to the direct and indirect effects of nuclear weapons. This vulnerability is real. It will remain a serious problem in spite of the ongoing efforts of both superpowers to protect and replicate critical components of their command and control systems. So destructive are nuclear weapons, however, that no amount of effort to make C^3I more mobile, redundant, or reconstitutable will ever suffice to make either superpower confident about its ability to function properly after being attacked. This goal is much more ambitious than is needed, however. All that deterrence really needs is to satisfy the limited and more readily attainable objective of guaranteeing retaliation.

The principal means by which the United States and the Soviet Union seek to safeguard their retaliatory capability is a redundancy in their command and control networks. This, coupled with some degree of predelegated launch authority, at least in the case of the United States, is meant to provide reasonable assurance of post-attack retaliation. In response to remaining uncertainties both superpowers have also developed quick-launch capabilities as another hedge against the possibly crippling effects of a nuclear attack.

Quick launch is a Soviet forte. The Soviets have always sought to deploy missile systems that are capable of quick or even automatic launch. They have gone to great lengths to develop the hardware and command structures that are essential to launch on warning (LOW) and launch under attack (LUA). Their Strategic Rocket Forces periodically rehearse these scenarios, making it likely that both do in fact reflect realistic Soviet options.[39] A string of statements by Soviet leaders indicates that this is at the least what Moscow would like the West to believe. Khrushchev, for example, declared shortly after the U-2 incident that Marshal Rodion Malinovsky, his defense minister, was empowered to respond instantly with a missile attack on any bases from which further U-2s might be sent.[40] The inference intended to be drawn from this and subsequent statements is that the Strategic Rocket Forces can launch missiles on their own authority in response to American attack.

American strategists, academic and official, give every indication of taking seriously the Soviet Union's putative capability to launch on warning.[41] Claims are also heard that Soviet leaders could continue to exercise control over their forces in the aftermath of an American attack because of their redundant, ground line–based C^3 system.[42] Most high-ranking U.S. military officials I have spoken to dismiss preemption as a suicidal option once the Soviets have brought their forces up to full wartime readiness.

Neither superpower would be prevented from retaliating if preemption took place after it had alerted its strategic forces. Nevertheless, many American strategists, in and out of government, believe that Moscow would still shoot first in a situation in which war looked highly probable. After all, the argument goes, the Soviets have had a preemptive strategy for decades and have invested considerable resources in developing the capability to carry it out.

But such a belief is not only dangerous in its implications, it is also quite likely wrong. Preemption may not appear appealing from the vantage point of Moscow. American strategic forces would presumably be fully readied for war in any acute crisis. In these circumstances Soviet leaders, military and civilian, would have little faith in their ability to prevent American retaliation.

Nor do we have reason to suppose that Soviet officials believe that the United States would really ride out a first strike. In the 1960s there was a discrepancy, revealed only later, between what the Kennedy and Johnson administrations said they would do in the event of a nuclear war and how they were actually planning to respond. The United States at the time was publicly committed to the doctrine of Mutual Assured Destruction and a countervalue targeting strategy. But at Robert McNamara's insistence, improvements were made in the strategic forces to give them a greater counterforce capability. Increasingly discrete counterforce options were also incorporated into the war plan (SIOP).[43] Soviet strategic analysts cite this discrepancy as an example of American strategic duplicity.[44]

Any conservative or suspicious Soviet planner—and they all

probably qualify on both counts—would have to assume that the well-publicized U.S. commitment to ride out a first strike had been enunciated for political reasons, domestic and foreign, and could not be taken as a guide to actual U.S. response to Soviet attack. To justify such cynicism, Soviet planners could draw not only on the McNamara precedent but also on a string of statements, from the time of General Curtis LeMay to the present, to the effect that SAC has never had any intention of following so passive a strategy.

LeMay was particularly outspoken on the subject of preemption. He was apparently committed to deciding on his own when and how the United States would fight a nuclear war.[45] To be sure, today's air force finds LeMay more of an embarrassment than an inspiration and does not question the president's authority to determine nuclear strategy. Nevertheless, air force generals still make statements that are at odds with official, or at least declared, policy. In May 1983 General Bennie Davis, then commander of SAC, told the Congress that it would be desirable if the United States could "ride out an attack rather than retaliating while under attack." But "as a practical matter," he insisted, "we have been unable to attain that [capability]" and must therefore make "a prompt response."[46] Davis's testimony implies that the United States would strike first if it became clear that the Soviets were preparing to attack, and it has the capability to do so.[47]

It is, of course, impossible to know just what the Soviets really think the United States would really do. Certainly the logic of the situation, statements like those of General Davis, and Soviet respect for American technology could all encourage belief in the existence of some kind of U.S. quick-launch option. To the extent that either superpower believes, rightly or wrongly, that the other is committed to launch on warning or launch under attack and would in fact execute such an option, much of its incentive for preemption disappears.[48] After all, little is to be gained from the destruction of empty missile silos. Alternatively, a Soviet analyst could conclude that the United States was likely to preempt at the onset of any Soviet strategic alert, a conclusion for which supporting statements can also be found. The apparent lack of a U.S. quick-launch option also encourages this conclusion, because such an option would be unnecessary if Washington were planning to strike first.[49] Perhaps this line of reasoning explains why Moscow was so careful not to alert any strategic forces during the Cuban missile crisis.

Recognition that preemption cannot forestall, or at least is highly unlikely to forestall, adversarial preemption does not entirely undercut its attraction. Preemption can appeal to policy makers because of their uncertainty about their own country's ability to retaliate. Doubts about this capability are made much worse by the pernicious phenomenon of worst-case analysis. "In war," Napoleon wrote, "one sees one's own troubles and not those of the enemy."[50] Contemporary military planners give every evidence of this kind of blindness. Because they are sensitive to their own weaknesses and their adversary's strengths, superpower strategists, more or less convinced of the other's ability to retaliate, remain doubtful, or at least uncertain, of their own. This parallel asymmetry is a powerful destabilizing factor in contemporary strategic relations.

Worst-case analysis also leads to exaggerated estimates of the adversary's military capability. In reality, a would-be attacker is painfully aware of everything that could go wrong; its leaders rightly worry about organizational and operational mishaps that could result in a poorly coordinated attack with weapons that do not reach their target or fail to detonate if they do. However, the would-be victim assumes that the adversary's preemptive strike will be an act of organizational and technical perfection, so devastating in its impact that it destroys the lion's share of its strategic forces and leaves the surviving remnants of its command structure in a state of shock and disarray, incapable of retaliation.[51]

Mutual, asymmetrical assessments of this kind greatly exacerbate fears of adversarial preemption. Neither superpower has confidence in its ability to prevent adversarial retaliation through preemption. But worst-case analysis brings both superpowers to believe that the other may be confident about its ability to do this. A preemptive strike at the height of a crisis consequently appears very likely. This expectation encourages both superpowers to consider "counter-preemption," if one can use the term, in order to protect themselves. Surely there can be little that is more absurd—or dangerous—than this reciprocal and self-fulfilling fear of adversarial preemption.

Damage Limitation

Analysts often argue that a preemptive, decapitating attack has real strategic advantages. By common consent, preemption is the only strategy that has any chance of actually winning a nuclear war. And failing victory, it might at least succeed in limiting damage to the aggressor. John Steinbruner, one of the most respected authorities on the organizational aspects of strategy, ventures this judgment:

> Unfortunately, a preemptive attack on the U.S. command structure is a rational defensive act for the Soviets once they have judged that nuclear war can no longer be avoided. Although it would preclude a bargained end of war, it offers two important advantages: First, by eliminating central coordination, it sharply reduces the military effectiveness of opposing strategic forces; second, it offers some small chance that complete decapitation will occur and no retaliation will follow. The latter possibility, however slight, is probably the only imaginable route to decisive victory in nuclear war.[52]

By Steinbruner's own reckoning, a decapitating attack at the height of a crisis offers only "some small chance" of forestalling retaliation.[53] The reason is that in a crisis the other side is fully alert to the possibility of attack and as a precaution against it has brought its own forces and command and control up to the highest state of readiness. In such a tense situation, officials at every level of command would probably be only too willing to give credence to reports that an attack against them was under way. Their ability to retaliate could also be enhanced by some kind of quick-launch capability.[54]

Decapitation, though it is theoretically possible, would not be so easy to achieve in practice. If the United States were the target, the Soviets would have to wipe out the primary and secondary command centers and their several airborne backups. Moscow would require a precisely coordinated lay-down pattern as well as advance information about the likely flight patterns of the Looking Glass and TACAMO aircraft. Even the destruction of these various command centers would probably not succeed in preventing retaliation, as submarine commanders unable to contact higher authorities in the aftermath of an attack reportedly have the authority to initiate retaliation on their own.[55] As there are no PALs on

SSBNs, loss of communication with the NCA would not constitute a technical obstacle to launch.

For all of these reasons, many nongovernmental analysts believe that some retaliation would occur in the aftermath of even the best-coordinated decapitating attack. In the words of Frank von Hippel: "What do Soviet leaders think U.S. submarine crews are going to do if they learn that the United States has been destroyed? Go to Tahiti and retire?" Just *one* Trident submarine is capable of firing twenty-four missiles, each armed with four 100-kiloton warheads—sufficient force to destroy the hearts of nearly one hundred Soviet cities, killing perhaps twenty to thirty million people.[56] Fifty Minuteman III missiles, launched by a still functioning missile wing, could deliver 150 warheads in the range of 335–350 kt. Presumably, these warheads would be pretargeted against important Soviet military assets, including command and control centers. As many of these targets are located in or near major urban centers, such an attack would also reap a considerable "bonus" destruction of people and industry.[57]

Steinbruner rightly concludes that a decapitating attack would be a frightful gamble, because the initiator whose attack fails to prevent an adversary from responding would provoke all-out retaliation. For the United States, the NCA functions as a "safety catch," to use Paul Bracken's metaphor, that holds back the multiple nuclear triggers of the unified and specified commands. A Soviet attack on the NCA would therefore constitute an attack against the safety catch of the entire command structure; it would destroy the only mechanism capable of preventing all-out retaliation. It would also render impossible any kind of subsequent effort to negotiate an end to hostilities and thereby limit destruction.[58]

A decapitating attack with the aim of preventing retaliation thus makes little sense if any chance exists of avoiding war. For no more than a minuscule chance of preventing retaliation, policy makers would make nuclear war a certainty and deprive themselves of any prospect of ending it short of the exhaustion of both sides. However, leaders in a crisis, subject to great stress, might not calculate the odds quite so rationally.

Psychologists have discovered important biases in the way in which human beings assess probability. If an outcome is perceived as disastrous but unavoidable, people are likely to reduce the probability they assign to the outcome, perhaps even to deny it altogether.[59] Most Californians conceive of earthquakes in this way; they go about their business pretending, even convinced, that the great quake which scientists predict is nearly certain to occur before the end of the century will never come to pass. Most policy makers do the same thing with regard to nuclear war.

When a particular outcome is believed to be the only possible way of avoiding great loss, by contrast, people tend to exaggerate its probability. Amos Tversky argues: "When it comes to taking risks for gains, people are conservative. They will take a sure gain over a probable gain." But, Tversky continues, "we are also finding that when people are faced with a choice between a small, certain loss and a large, probable loss, they will gamble."[60] This phenomenon has been well-documented in foreign policy decisions; Chapter 4 will describe how it has been an important cause of miscalculations that led to undesired wars.[61]

Which bias would dominate in a nuclear crisis? Policy makers might downplay the risk of war because of their knowledge of its destructiveness—a form of denial that could have other adverse consequences. But it is also conceivable that Soviet or American leaders, at the height of a crisis that seemed likely to spill over into nuclear war, might grasp at the straw of preemption precisely because preemption seemed to offer the only means by which they might protect themselves and their country. For this reason, it is imperative to carry out and publicize more objective assessments of the trade-offs involved in preemption *before* a crisis develops. We need such assessments as an antidote to the wishful thinking induced by stress.

Many of these arguments also apply to another alleged appeal of preemption: that it would sharply reduce the military capability of the adversary. The claim is indisputable; it is also meaningless. Blunting an adversary's offensive arsenal could in theory confer two kinds of advantage. It would limit the damage that subsequent retaliation inflicted on the initiator, and it might permit the aggressor to end the war on terms more favorable than would otherwise be the case.[62]

U.S. analysts have calculated that preemption would significantly limit the number of casualties that the attacker would sustain. It is generally assumed that in a major nuclear war the United States would suffer somewhere between 155 and 165 million near-term fatalities. Fifty million lives, so the argument goes, might be saved by striking first. For the USSR, the disparity is even greater. The Soviets could be expected to suffer about 120 million fatalities as a result of full implementation of the U.S. SIOP but only 50 million if they struck first. The difference is seventy million lives.[63]

But these figures are misleading. Even if they are accurate, they represent only the fatalities expected from blast, radiation, and fire in the first week. Many more millions would die in the course of the following year because of the collapse of medical care, food production and distribution, public health, and other essential services. The number of survivors one year later might not be much different regardless of how many people actually lived through the first week. It is also unclear whether society as we know it could survive the cumulative effects of this kind of shock and disruption.[64] Finally, there is the threat of nuclear winter; if the more pessimistic studies are correct, the number of immediate survivors will be a meaningless statistic.

Despite these discomforting realities, strategists in both superpowers give every indication of believing that preemption promises to confer a meaningful postattack advantage. Michael MccGwire argues that the difference between 120 and 50 million dead is especially important to the Soviets because of the trauma of World War II. At the same time that experience has also made them even more cautious of war, a concern that presumably militates against preemption if any chance remains of preventing war.[65] After all, no fatalities are even better than fifty million. These somewhat contradictory objectives, MccGwire argues, have resulted in a strong commitment in Moscow to do everything possible to avoid war. But if war nevertheless breaks out, Moscow is committed to use force massively in order to limit damage to itself.

In the United States official strategists also conceive of political and military costs in relative terms.[66] The nature and magnitude of the costs are not considered in and of themselves; they take on meaning only in comparison to expected gains. Such a theoretical formulation ignores the reality that for political leaders, *absolute* costs, when sufficiently great, are a very important consideration. McGeorge Bundy, special assistant to the president for national security affairs in the Kennedy and Johnson administrations, has criticized strategic analysts for their lack of realism in this regard:

> There is an enormous gulf between what political leaders really think about nuclear weapons and what is assumed in complex calculations of relative "advantage" in simulated strategic warfare. Think tank analysts can set levels of "acceptable" damage well up in the tens of millions of lives. They can assume that the loss of dozens of great cities is somehow a real choice for sane men. They are in an unreal world. In the real world of real political leaders—whether here or in the Soviet Union—a decision that would bring even one hydrogen bomb on one city of one's own country would be recognized in advance as a catastrophic blunder; ten bombs on ten cities would be a disaster beyond history; and a hundred bombs on a hundred cities are unthinkable. Yet this unthinkable level of human incineration is the least that could be expected by either side in response to any first strike in the next ten years, no matter what happens to weapons systems in the meantime.[67]

The size of superpower strategic arsenals and the near certainty that some kind of retaliation would occur make meaningful damage limitation an illusory notion. Even a small number of surviving delivery systems would be sufficient to inflict horrendous punishment. The greater concentration of Soviet population and industry in and around several score urban centers makes it even more vulnerable to destruction by a "modest" attack. In the 1960s McNamara's defense analysts calculated that only about 50 equivalent megatons would be required to destroy more than half of the Soviet population and economic base.[68] Admittedly, the weapons of that period were of larger yields than their present-day counterparts, so more weapons would now be needed to inflict the same amount of countervalue destruction. They would still represent a very small proportion of either superpower's nuclear capability, and at least this much in the way of warheads could reasonably be expected to survive any attack.

Preemption could conceivably result in *more* rather than less damage to the initiator. In the hope of forestalling retaliation, the attacker would target the adversary's political and military leadership as well as his strategic forces. As so many of these targets are within or near to major urban areas, attacks against them would cause tremendous collateral damage. This would invite all-out retaliation, much of it deliberately directed against the initiator's own population and industry. If preemption failed to prevent retaliation, the surviving strategic forces could, and presumably would, carry out truly devastating attacks in retaliation. Nor would it be easy, if it were possible at all, to negotiate a subsequent cease-fire or end to the war in the absence of an intact political leadership on both sides.

Military Advantage

Could preemption secure a military advantage that could be translated into more favorable peace terms? Contemporary Soviet and American strategy appears to be based on this premise. The American "countervailing strategy" assumes that a nuclear war can be won and that victory will go to the side that has the greatest residual military capability, nuclear and conventional. As a result, a survivable counterforce capability is deemed essential for deterrence. But such an approach to nuclear war, as Robert Jervis has demonstrated, is based on erroneous political and technical assumptions.[69] Three of these are particularly germane to any assessment of the utility of preemption.

The first, and by far the most questionable, is the belief that there could be a winner in a superpower nuclear war. The authors of the 1954 American war plan believed that an American attack would render the Soviet Union a blackened, radiating ruin.[70] Today this description would apply to both superpowers. Without leaders, their economies and societies in ruins, and most of their people dead or dying, the United States and the USSR would cease to be superpowers. Their relative surviving military strength, if any, would be overshadowed by that of undamaged third parties. Military strength, in any case, would not be an important determinant of post-attack economic recovery.[71] If the war produced a nuclear winter, then the survival of both countries, perhaps of the entire human race, would be in jeopardy.[72]

A second point to consider is how the superpowers could determine their relative military capability after the fighting stopped. With this end in mind, the U.S. is planning to deploy the Integrated Operational Nuclear Detection System. Expected to be operational by 1988, it will place nuclear explosion detectors on eighteen NAVSTAR satellites. This Nuclear Detection System is designed to pinpoint and measure nuclear explosions anywhere in the world and thus to provide real-time damage assessment information to U.S. strategic forces. NAVSTAR satellites will not depend upon ground-based relay stations; they will be able to transmit detection system data direct to airborne command posts.[73] However, this capability may well be for naught; it is unlikely that NEACP or Looking Glass would survive for very long in a postattack environment.[74]

The absence of functioning communication facilities or of political leaders with access to them would make it impossible to ascertain which, if any, of one's own forces were still intact and operational. These forces would in any case be useless for bargaining purposes unless communications could be maintained with them. The strategic forces most likely to survive are SSBNs. Ironically, they would also be the element of the strategic triad most difficult to communicate with in a post-attack environment. Very-low- and low-frequency antennae, satellite ground stations, links to TACAMO aircraft, and the TACAMO aircraft themselves would be primary targets of any Soviet strike. Any component of the NCA fortunate enough to survive the opening salvo of a nuclear war would probably find it impossible to contact the submarine force, let alone command it in any meaningful way.[75]

Without satellite surveillance it would be even more difficult than it is now to form a judgment about the adversary's military capabilities. Even if photographic and signals intelligence satellites were still in orbit and functioning, their few ground stations and data analysis centers would almost certainly have been blown away.[76] In effect, both superpowers would be blind. They would be unable to procure good information about either their own military capabilities or those of the adversary. And so relative residual military capability would be not only irrelevant but also unknowable. Even if one superpower knew something about its own scattered military forces, any advantage would be useless for bargaining purposes unless its adversary had some means of verifying that knowledge independently. The only other way one superpower could convince its adversary of remaining military prowess would be to use it—but military might then would lose its political value as a form of leverage.

Of course, this entire discussion begs the question of whether any national command authority would be functioning in the aftermath of a nuclear exchange in which one or both sides specifically sought to destroy and disrupt as much of the adversary's command and control as possible. And this observation brings us to the third point: intrawar or postwar bargaining requires responsible political leadership on *both* sides. Without leaders, or in the absence of any effective means by which the two leaderships could communicate with each other and with their surviving forces, relative residual military capability would be meaningless. This requirement highlights another irony of war-fighting strategies. The conditions essential for exploiting a post-attack military advantage would prevail only in a "limited" nuclear war in which each side deliberately refrained from attacking the other's political leadership and C^3I. By definition, this would rule out a decapitating attack.[77]

Technology versus Politics

It is apparent that there is little political logic to a decapitating attack or, for that matter, to a preemptive strike of any kind. How then can we account for the widespread notion in the strategic literature that it is rational to consider preemption in a crisis that goes to the brink of war? One explanation is the failure of many strategists to consider carefully the likely consequences of preemption. Another is the hold that technology, independent of any political consideration, exercises on the minds of so many American strategic analysts and defense officials. C^3I is a tremendous force multiplier but also extremely vulnerable, a combination that naturally leads to a desire to exploit the capability before it is destroyed or unusable. The pressure to preempt is therefore the result of a commitment to use the military potential of one's strategic forces to the fullest *independently* of judgments of political utility. In a complete reversal of Clausewitzian logic, technology, divorced from politics, has come to dominate strategy.

It is remarkable how little contemporary strategists have learned from the experience of 1914. Although strategists the world over pay lip service to the major military lesson of the July crisis—that strategy must be subordinate to policy—they have helped violate this cardinal principle with respect to nuclear weapons. The reasons why this has happened are also reminiscent of 1914.

The Schlieffen Plan made politics subservient to strategy in order to capitalize upon Germany's military advantages: a dense rail network and a national penchant for military organization. The German generals envisaged the Schlieffen Plan as the only possible strategy by which they could win a European war. The general staff embraced the plan for this reason, even though they recognized it constituted a great gamble. By 1914 many high-ranking officers, Moltke among them, were pessimistic about the likelihood of the plan's success. They stuck to it because they saw no alternative.[78]

The German military also knew that the Schlieffen Plan made war more difficult to avoid in a crisis. Yet over the years they invested in the manpower and infrastructure essential to their strategy. This investment became another incentive for implementing the strategy—and to do so preemptively, because the military preparations of their adversaries made it increasingly likely that only with German preemption would the Schlieffen Plan work. Not only did the Schlieffen Plan increase the probability of war, it also guaranteed that the resulting struggle would become total and unrestrained. Today's war-fighting strategies have a similar effect.

There is a second disturbing parallel to 1914. Preemption occurred in 1914 because of the almost universal belief that victory would go to the side that mobilized first. We know now that this was an egregious illusion. Prompt mobilization could not guarantee military victory, nor would delayed mobilization have caused defeat. This misunderstanding had tragic consequences: it stampeded policy makers into a war that they could have avoided. A similar misunderstanding may exist today. Many contemporary American strategists appear to overvalue the military utility of preemption and undervalue its political and material costs. The reason for doing so is the same as it was in 1914: a failure to consider the *political* context in which weapons would be used. Unless this myopia is corrected, the illusory fear of loss could once again prompt preemption in a crisis in which otherwise a diplomatic solution might be found. The consequences of preemption, horrendous as they were in 1914, would pale in comparison to what we might expect from a superpower nuclear war.

When Is the Inevitable Inevitable?

Most discussions of preemption take for granted that both superpowers want to avoid war at almost any cost. They would contemplate preemption only in a situation where war seemed unavoidable. If this assumption is correct, the important question becomes how the superpowers would conclude that war was unavoidable. What political or military indicators of war would leaders depend upon? How reliable would they be?

Three kinds of conflict immediately come to mind: a conventional war in Europe, a crisis that appears to be spinning out of control, and an advance warning to one of the superpowers that the other intends to attack. Each of these situations might be considered a prelude to nuclear war. But each is also certain to be highly ambiguous.

A conventional war in Europe is likely but by no means certain to escalate into a nuclear conflict. Many experts believe that it could be contained at the conventional level.[79] Others argue that containment would be extremely difficult. Some authorities believe that it might be possible to keep a theater nuclear war from escalating. Others insist that once nuclear weapons are used, the conflict will escalate rapidly to the intercontinental level. One reason they give is the colocation of so many Soviet conventional and strategic assets, whose potential destruction would generate an enormous pressure to preempt.[80]

Opinion is also divided about the implications both of a superpower naval encounter and of combat between Soviet and American ground forces somewhere other than in Europe. Desmond Ball argues that a naval battle could easily involve nuclear weapons.[81] But at the same time naval engagements may be easier than land battles to isolate and control. An incident at sea could prompt orders from above for rapid disengagement because of fear of all-out nuclear war it would arouse on both sides.

All of these scenarios are surrounded by uncertainty; informed analysts disagree about the chances of escalation into intercontinental nuclear war and the causal mechanisms that would be responsible should that happen. Much of the disagreement can be attributed to the difficulty of determining in advance just how much loss, or threat of loss, leaders of either superpower would be willing to tolerate in order to keep the bogey of nuclear war at bay. If *both* were to exercise restraint, then a conventional encounter would not develop into nuclear war (assuming, of course, that leaders were able to maintain effective control over their own forces). But if either superpower concluded that the other was about to strike at its homeland with nuclear weapons, then it might deem it essential to get in the first blow. But, as we shall see, the judgment that the other superpower is about to strike can never be made with full certainty. Accordingly, the side that strikes first risks making its fear of nuclear war unnecessarily self-fulfilling.[82]

Another possible route to war is a crisis in which one of the superpowers concludes that nuclear war is unavoidable because of loss of control. Some strategic analysts contend that high levels of military alert could readily bring such a situation about.[83] The next chapter explores the mechanisms by which loss of control could result in war. Grant for the sake of the argument here that a nuclear war could arise in this manner: Would preemption be a feasible option?

Loss of control and preemption posit quite opposite conditions. We would expect the former to result from a breakdown in central control over nuclear forces; the latter presupposes effective central control and coordination of those forces. For leaders to preempt in response to loss of control, they would have to recognize that they were losing their ability to control their forces but still possess enough control to launch a coordinated first strike. Alternatively, they would have somehow to learn that their adversary was about to lose control and choose to preempt in the hope of limiting damage to themselves. It is not easy to envisage either situation happening in practice.

Policy makers might not realize that loss of control had occurred until the first weapons had been fired or until after it was too late for them to coordinate a first strike. If they recognized the problem, they would probably be in a position to prevent war by establishing tighter control over their forces. Presumably they would choose to do this in preference to attacking. Loss of control could also occur as the result of a major communications failure that severed the links between the National Command Authority and components of the nuclear forces. But in the case of such a failure, even a president or a premier who wanted to would probably be unable to order or coordinate a preemptive strike.

Although war arising from loss of control is a real possibility in a serious crisis, preemption is unlikely to be the mechanism to bring it about. If it did, it would probably be because leaders panicked; afraid of the adverse military consequences of loss of control, they could decide to preempt while they still had the capability to do so. Loss of control could be a catalyst of war in this situation, but its cause would be all of the largely irrational or exaggerated fears that I have already described.

A third situation in which policy makers might conclude that nuclear war was unavoidable is in response to warning of attack. The warning could be strategic, that is, well in advance of any attack. It could be conveyed by a highly placed source in the adversary's government, or it could be the result of communications or signals intelligence. The classic example of the former is the 1941 report by Richard Sorge, the Soviet spy in Tokyo, that Japan had decided not to attack the Soviet Union but to move instead against Southeast Asia.[84] Stalin made good use of this information: he shifted forces that had been guarding the eastern frontiers to the west, where they proved vital to the defense of Moscow. The Americans, by contrast, broke the Japanese naval code well in advance of December 1941 and intercepted word of the impending attack against Pearl Harbor. They nevertheless failed to profit from the advance warning.[85]

Strategic warnings are godsends, but they are not always correct. The spy can be mistaken or deceived, or form a judgment on the basis of incomplete information.[86] Signal intelligence can also be misleading or misinterpreted. More to the point, strategic warning may simply not be relevant to the question of superpower nuclear war. It presupposes that a decision to go to war will be made some time before any attack. Adversaries have some prospect in the interim of learning about the decision through various means of intelligence. They could get wind of it from one of the participants in the deliberations, from an indiscretion of someone around them, or from information communicated to other officials involved in planning the attack. None of this is likely to happen in the case of superpower nuclear war.

Both sides expect that nuclear war would be catastrophic, and so it seems almost inconceivable that the leaders of either superpower would cold-bloodedly make a decision to go to war at some future date.[87] Neither side is going to opt for war if it believes any chance exists to prevent it. War, if it comes, will be the result of a crisis that either spins out of control or becomes so acute that one of the superpowers preempts for fear that the other is about to do so. In such circumstances the decision for war, if indeed there is one, will be taken by a small circle of top leaders after they have already put their strategic forces on full alert. It will also be executed as promptly as possible, probably in a matter of minutes. Advance warning is unlikely. If any is received, it will arrive at about the same time as tactical warning.

Strategic warning also includes the variety of preparations that observers expect to precede an attack, measures not normally associated with peacetime activities. In the case of nuclear attack such preparations would include bringing strategic forces up to the highest state of readiness, sending bombers aloft and dispersing others to secondary airfields, and ordering submarines still in port to sea. Meanwhile, surface and underwater vessels would try to avoid Soviet surveillance, secondary and tertiary airborne command posts would be activated, and satellite and airborne reconnaissance of the adversary's forces and their military preparations would intensify.[88]

Most if not all of these measures would be detected by the other side. But they are by no means unambiguous indicators of attack.[89] They could just as easily be defensive measures, implemented in response to fears that the adversary was preparing to attack. Strategic indicators of attack would be more revealing in the absence of a crisis, but in such a circumstance preparations for nuclear war also seem highly unlikely. They would still not constitute certain proof of impending attack.

The ambiguity of warnings means that policy makers can never be really certain that the adversary is committed to war until the adversary actually launches weapons. Even then, as the next chapter will argue, there is an element of uncertainty. To the extent that policy makers or their advisers recognize this uncertainty, any decision they make to preempt on the basis of strategic warning would be influenced not only by a belief in near inevitability of war but by other considerations as well. The most important of these is probably the belief, however erroneous, that striking first confers significant advantages. When a superpower exaggerates the other's preparations for war and overvalues the advantage of striking first, it makes preemption more likely to occur.

The inescapable conclusion is that preemption is an altogether irrational act. For the prospect of little or no real gain, it risks unleashing an unnecessary nuclear war. Its irrationality does not preclude it from happening, of course; history is full of examples of irrational, self-defeating, and destructive behavior. Preemption, if it does occur, will almost certainly be an emotional overreaction to acute threat. It will be triggered off by fear of the other side's intentions and fear of the consequences of not preempting. Both fears could be greatly exaggerated, as they were in 1914. The history of that crisis suggests a sobering conclusion—prevention of a modern-day tragedy requires a major effort, well in advance of any future crisis, to dampen the role that we can expect exaggerated threat assessments to play during a crisis. A useful start in this direction would be a more thorough exposure of the false strategic assumptions on which these assessments are based.

Assessing the Risks

The vulnerability of strategic command and control could have catastrophic consequences quite different from those envisaged by the American military. An escalating spiral of defensive precautions could make superpower fears of a crisis leading to war self-fulfilling. This threat is intensified by three other likely attributes of any future superpower crisis: superpower lack of familiarity with each other's alert procedures; the tendency to apply worst-case analysis to military intelligence; and the American penchant for relying upon military actions to signal resolve. These factors would aggravate mutual fears of attack.

The Meaning of Strategic Alerts

Prior to 1914 no great power had mobilized since 1878 except for Russia, during the Russo-Japanese War. And that mobilization was directed toward the Far East; it had little impact upon European understanding of military affairs. A. J. P. Taylor suggests that lack of experience with mobilization made European general staffs reluctant to tamper with any of their war plans. At the same time it reinforced their belief that states did not mobilize for any purpose other than war.[90] The Austro-Hungarian and Russian mobilizations in 1914 accordingly aroused widespread expectations of war. These expectations were important causes of war in their own right.[91]

The shock of mobilization could have similar consequences today. Neither alliance has ever carried out anything approaching a full-scale mobilization. Preparations of this kind, coming in a crisis, could tip the balance in favor of war even if intended as a deterrent. After all, the Russians conceived of mobilization in 1914 as a deterrent. Instead, Russian mobilization prompted German mobilization and war, because the Germans believed that failure to act would put them at a significant military disadvantage. Today, full or partial Warsaw Pact mobilization would prompt a NATO response. Like Germany before it, the Soviet Union would be hard-pressed to go to war before NATO mobilization progressed to the point where it deprived the Soviets of a meaningful numerical edge.[92]

The military situation is also asymmetrical with respect to strategic weaponry. American strategic forces are always kept at a higher state of readiness than the Soviets' and have on several occasions been brought up to even higher levels of alert. In 1962, during the Cuban missile crisis, President Kennedy put American forces, conventional and nuclear, on a near-war footing. Again in 1973, American forces were put on alert in response to a Soviet threat to intervene in the Middle East: they were brought to Defense Condition (DefCon) III, the second highest state of readiness short of war.[93] According to Henry Kissinger, DefCon III "is in practice the highest state of readiness for essentially peacetime conditions." The National Security Council was even prepared to discuss additional alert measures, but they proved unnecessary as the threat of Soviet intervention abated.[94] On neither occasion were Soviet strategic forces put on alert.

Until the late 1970s the Soviet Union kept its forces at a very low state of readiness.[95] Western analysts speculated that this was the result of several reinforcing considerations, among them the technical characteristics of Soviet forces, the leadership's desire to maintain tight control over nuclear weapons, and the apparent Soviet belief that war is very unlikely to begin with a "bolt from the blue" attack. Rather, the Soviets expect it to break out after a period of great tension. There will have been ample time to bring strategic forces up to wartime readiness.[96] Today a greater percentage of Soviet forces is maintained at operational readiness. Although this percentage is still significantly lower than that of the United States, it provides the Soviets with roughly the same capability in deliverable warheads.

The Soviet Union also conceives of the political utility of its strategic forces quite differently from the United States. In neither an exercise nor a crisis has it ever put these forces on alert. During both the Cuban missile crisis and the Yom Kippur War, Soviet strategic forces reportedly remained at their normal state of readiness, which at that time was extremely low.[97] In 1973 Moscow did place some naval and ground forces on alert, among them its naval squadron in the Mediterranean and all seven airborne regiments inside the Soviet Union. These actions were probably designed to lend credibility to the threat to intervene.[98] Western analysts have speculated that Moscow failed to put any strategic forces on alert in either the Cuban or the Middle East crisis because the Soviets wanted to avoid the kind of provocative action which might have significantly increased the risk of war with the United States.[99]

The fact that the Soviets keep their strategic forces at a relatively lower state of alert, and have never changed their alert status in a crisis, would make any decision by Moscow to bring these forces up to wartime readiness all the more threatening. American leaders could interpret such action only as an indication that Moscow believed war was a real possibility. Would such a move have a deterrent effect? Or would it trigger war, as the Russian mobilization of 1914 did?

A Soviet strategic alert would unquestionably convey a heightened sense of threat to American leaders. It might cause them to rethink their policy and to search more single-mindedly for a way out of the crisis. But it would also aggravate any American propensity to see war as unavoidable, and it could generate strong pressures to preempt *before* Soviet forces reached full wartime readiness.[100] One can imagine a "hawkish" presidential adviser in the White House situation room urging preemption as the only possible course of action. The adviser would contrast Soviet strategic preparations in the current crisis with Soviet passivity in previous crises and argue that Moscow would not now be generating its forces unless it expected to use them. The United States, this adviser might insist, must seize the initiative before a Soviet first strike decimated its forces and their command and control. Soviet force generation makes war all but unavoidable, the argument might proceed, because both sides would face such great difficulties in maintaining control over their respective strategic forces once they had brought them up to the highest levels of alert. We can only hope that no president would be swayed by these arguments.

Exaggerated Threat Assessments

In 1914 the sequential mobilizations of the great powers were premised on the belief that one or more adversaries had already begun to mobilize secretly. The Russians thought that Austria-Hungary was surreptitiously mobilizing against Serbia. The Germans became convinced that Russia was doing the same against them, while the French believed that Germany was getting ready to invade before their army was in a position to defend itself. The general staffs of all three countries, Russia, Germany, and France, argued that secret preparations of their adversaries were proof of aggressive designs and of the inevitability of war. They warned political leaders that a failure to mobilize would give the other side a decisive advantage. Conrad and Moltke on the one side, Ianuskevich and Joffre on the other, pushed their respective governments into making premature decisions. Fears of war became self-fulfilling.[101]

We know today that all of the several general staffs were mistaken about their adversaries' secret military preparations. Research in the archives of the countries involved indicates two causes for this tragic illusion. Military intelligence organizations in all the Continental powers overreacted to ambiguous information and unsubstantiated rumor. Their exaggerated assessments were not subjected to any independent scrutiny by military or political leaders. Misjudgment also stemmed from each country's lack of familiarity with the mobilization procedures of its adversaries. The details of mobilization were kept secret, something helped by the fact that nobody had mobilized, or even rehearsed mobilization, in Europe since 1878. Russia and Germany, as noted, both seem to have mistaken premobilization preparations for secret mobilization and to have drawn obvious but incorrect conclusions. Fear too must have played its part. As the generals worried about their adversaries getting a jump on them, so they became overly receptive to information that suggested this was actually taking place.[102]

Contemporary intelligence organizations are more sophisticated in their collection and analysis of data. They are unlikely to venture a firm judgment about an adversary's military preparations on the basis of a few wisps of ambiguous or uncertain information. But they are just as likely to misjudge the extent of their adversary's military preparations in a crisis because they are equally unfamiliar with the routines that are involved. Each side regularly monitors the other's strategic exercises, trying to put together a picture of the steps involved in getting up to wartime readiness. But exercises are not the same as the real thing. As the Soviet Union has never put its forces on high alert, and the United States has never gone beyond DefCon III across the board, both sides must be working with an understanding that is necessarily far from complete.

Monitoring of escalation in a crisis would also be hindered by inadequate, incomplete, and even deliberately misleading information. Either superpower might try to mask the extent and intensity of its military preparations; the Soviets in particular place a great emphasis on strategic deception.[103] Washington or Moscow might also attempt to spoof or jam the other's sensors as part of this deception. Cloud cover could fortuitously interfere with visual reconnaissance. In these circumstances it would be easy to exaggerate the extent of an adversary's preparations. Exaggeration would be all the more likely if leaders and intelligence officials expected their adversary to engage in strategic deception. American officials certainly harbor this expectation, and they would accordingly be inclined to believe that more military preparations than what they were able to detect were under way. Any evidence or even suspicion of jamming or spoofing would also convey a heightened sense of threat.

The deteriorating situation in Poland in the winter of 1980 illustrates how incomplete information can encourage an exaggerated assessment of threat. The White House on 3 December began to express alarm about the possibility of a Soviet invasion. On 7 December it announced that the Soviets had in fact completed preparations for an invasion.[104] Early in the new year, however, Les Aspin, chairman of the House Intelligence Oversight Committee, said the Carter administration, drawing conclusions from intelligence estimates, "grossly exaggerated the state of readiness of the Red Army. . . . Contrary to the impression given in most news stories, the Kremlin is still getting its ducks in line."[105] Subsequent reports indicated that cloud cover caused problems for satellite observation in December 1980 and January 1981, contributing to the unfounded reports of the CIA.[106]

Anticipation of a Soviet intervention in Poland was pronounced in Washington that winter. This atmosphere, one can surmise, made the intelligence community particularly receptive to information that pointed to an invasion. Analysts may have jumped to the conclusion that the presence of a military command post indicated an associated deployment of forces, even though they could not actually detect such forces. What forces they did locate could have been taken as evidence for the presence of even more, unseen troops. Indeed, just such a buildup had preceded the invasion of Afghanistan the year before; it was hardly surprising that intelligence analysts assumed it now presaged an invasion of Poland.[107]

The phenomonen of cognitive consistency—of seeing what you expect to see—could just as readily interfere with intelligence assessments in a direct confrontation between the superpowers. If a crisis were preceded by a period of rising tensions and fears or, worse still, expectations that the other side was preparing to resort to force, analysts would be on the lookout for evidence that confirmed these expectations. And, no doubt, if they looked hard enough, they would find it.

Actions versus Words

The third and probably the most dangerous structural attribute that would encourage preemption is the American belief that actions signal intentions better than words. The Soviets appear to be much more circumspect in this regard.

Stephen S. Kaplan reports that the Soviet Union used its armed forces as a political instrument on 190 occasions between June 1944 and August 1979.[108] These include instances in which force was actually used, as in the invasions of Hungary and Czechoslovakia, as well as situations where the Soviets intended to convey a political message through conventional force alerts, naval deployments, or the sending of military advisers. The great majority of these actions were directed against third countries. Moscow has been chary of using its military might as a means of sending signals to the United States. The obvious exception is October 1973. Then the

Soviet threat to intervene in the Middle East, backed up by an alert of airborne forces, was seemingly intended to communicate to Washington that the Soviets felt they had important interests at stake.

The United States, according to a parallel calculation, flexed its military muscle 215 times between 1946 and 1975.[109] Many of these actions were attempts to send political signals to the Soviet Union, a pattern that has continued, if not intensified, since 1975. Although the United States may not use the military instrument any more than the Soviet Union, it does use it in different ways. American leaders are much more likely than their Soviet counterparts to resort to military preparations or presence as a means of communicating resolve, especially in crises. American leaders are also more likely to direct these signals toward their principal adversary.

This American proclivity for relying upon military means to signal resolve is a matter we can probably explain in terms of several mutually reinforcing considerations. The first of these is technical. Air and naval forces are more flexible, mobile, and controllable than ground forces and hence more suitable for sending political messages. Naval forces do not require bases in the immediate vicinity of where their presence is required. The United States commands an impressive array of naval and air resources that can be used for political purposes; carrier task force groups, which combine naval and air platforms, are particularly well-suited to this task. The Soviet Union, by contrast, is at a significant disadvantage despite the ongoing expansion of its naval arm.

The military services also have a strong institutional incentive for encouraging political leaders to use military forces as a political instrument. The routine dispatch of carrier groups to trouble spots around the world provides an important justification for the navy's claim that its aircraft carriers and their large number of supporting vessels are essential to national security. Aircraft carriers participated in 106 of the 215 incidents investigated by Barry M. Blechman and Stephen M. Kaplan, a level of involvement that the navy and marine corps are fond of pointing out to the Congress at appropriation time.[110]

Finally, the American penchant for military displays reflects the nation's postwar fixation about credibility. The widely shared belief that only resolve deters aggression has as its corollary that the surest way of demonstrating resolve is frequent demonstrations of willingness to use force in defense of one's commitments. This may be why, as Richard Betts has documented, political leaders have been more prone than military officials to favor the use of force in crises and other serious confrontations.[111]

Military displays have accordingly become standard operating procedure in time of crisis. When a president confronts a foreign threat, he reaches almost by reflex for a carrier group or a fighter squadron to send to the troubled area. Between crises, Pentagon planners prepare military options for the White House, almost all built around demonstrations of military muscle. Options of this kind also constitute the principal moves in the simulations of Soviet-American confrontation in which high-ranking political and military officials sometimes participate. Having written and participated in such simulations, I know from personal experience that political leaders are even more inclined than their military counterparts to seize upon military options as a way of displaying resolve.[112] These exericses, in which obviously no real Soviets ever take part, tautologically confirm for policy makers and their advisers the utility of military displays. Academic studies, some of which raise serious questions about the utility of these tactics, do not seem to have made any impact on the way in which policy makers shape their responses to crises.[113]

Gunboat diplomacy may be the accepted American way of dealing with adversaries, but it is a dangerous atavism in the late twentieth century. For very little in the way of apparent political return, it raises the risk of war. The Cuban missile crisis and the Middle East war of 1973, often cited as successful examples of crisis management, are in fact inappropriate models for contemporary Soviet-American relations. These confrontations occurred at a time when the alert and response systems of both superpowers were less mature and less tightly coupled than they are today. Strategic escalation in both crises was, moreover, an entirely one-sided affair. As a result, it risked few of the dangers that future crises will entail. Mutual escalation today, as I demonstrate in subsequent chapters, would raise the specters of loss of control and miscalculated escalation in addition to that of preemption.

Not only may crisis escalation be dangerous, it may often be unnecessary. Its avowed purpose from the American perspective is to convince Soviet leaders that Americans view the interests at stake as vital. But the record of past crises offers very little support for the conventional wisdom that states evaluate one another's resolve on the basis of efforts to impart credibility to commitments. Leaders actually appear to be much more sensitive than analysts have assumed to the nature and gravity of the interests they judge to be at stake. When vital interests are on the line, demonstrations of resolve are often unnecessary. Or they fail to convince—as so often occurs when leaders of a country commit themselves to a provocative course of action.[114]

Displays of military prowess or preparedness may, in fact, be most instrumental in influencing how the adversary calculates resolve in confrontations where interests perceived as marginal are at stake. The American threat to use nuclear weapons in defense of the offshore Chinese islands of Quemoy and Matsu in 1958 is a case in point. The threat, to the extent it was believed in Peking, may have influenced the People's Republic to exercise caution in the confrontation. However, threats of this kind are probably less efficacious when made against nuclear adversaries. If the People's Republic in 1958 had had its own nuclear arsenal and the means of delivering warheads, the American threat to use nuclear weapons would have entailed much greater risk and for this reason would have been much less credible. More to the point, it would have risked a nuclear war over an issue that, though it was deemed important at the time, was certainly never seen as vital to American survival. The lesson is clear: if the interests at stake are tenuous, their defense cannot be worth serious risk of nuclear war. For the same reason, threats to resort to nuclear war in defense of them are not likely to be believed.

This is not to suggest that demonstrations of resolve lack all utility. Clearly a range of situations exists in which they can abet the goals of statecraft. It seems unlikely, for example, that Khrushchev would have agreed to withdraw Soviet missiles from Cuba had Kennedy not imposed the blockade and subsequently threatened an airstrike. At the same time the president's actions could have provoked a war. Fortunately for Kennedy, his coercive policy achieved its desired goal. Before resorting to policies of this kind, statesmen must carefully weigh the risks involved. But even if they do, recent history indicates, there is ample room for miscalculation.

The fundamental cause of crisis instability today is the marriage of the most modern weapons of destruction to antiquated conceptions of conflict management. These conceptions, even more than the weapons themselves, threaten in the future to push any acute crisis toward war. They make policy makers and their advisers unduly sensitive to threats that are unrealistic or farfetched, and they make them correspondingly insensitive to other threats that may be more likely and graver. And so the very focus of contemporary strategic concerns is misplaced. This will become even more apparent in the next chapter, which takes up the problem of control.

[3]

Loss of Control

In this chapter I take up our second sequence to war, loss of control, and I describe several different ways in which it could lead to unintended nuclear war between the superpowers. I identify the likely causes of each of these paths of war and show the extent to which they are structural attributes of the superpowers' alert and response systems. A high level of risk is, I shall show, inherent in strategic force generation.

Loss of control is analytically distinct from preemption. In neither sequence is war a desired outcome, but in the case of preemption war nevertheless results from a deliberate decision by leaders on at least one side to wage it. When loss of control leads to war, by contrast, it does so because of the actions of subordinates which leaders are unaware of or unable to prevent.

Loss of control can nevertheless be a contributing cause of preemption. If, in a crisis, leaders become convinced that war is inevitable, because they can no longer maintain control over their forces, they may decide to use them preemptively if they think preemption will confer a significant military advantage. Loss of control is also linked to miscalculated escalation, but here the sequence is reversed. Miscalculated escalation refers to steps up the political-military escalation ladder in a crisis, steps taken to moderate adversarial behavior which instead provoke further escalation by the adversary. It can thus lead to war by loss of control. This chapter illustrates the dynamics of this process in the context of strategic alerts.

Loss of control can have political or institutional causes. In the former case disaffected officials try to sabotage policy or impose a new policy of their own; this was an important cause of war in 1914. Institutional loss of control is a more complex phenomenon. It occurs when individuals, acting on orders, or at least within the accepted confines of their authority, nevertheless behave in ways that interfere with, undercut, or are contrary to the objectives that national leaders are pursuing. Institutional loss of control arises because policy decisions in large bureaucracies often have significant unanticipated consequences.

Military Rigidity

Modern political and military organizations are complex bureaucracies. Their characteristic modes of operation can be quite inappropriate to the needs of political leaders in a crisis. Crisis management requires political finesse in the formulation of policy and a surgical precision in its execution. Generally, large bureaucracies are incapable of either. It is for this reason that political leaders frequently rely on a small coterie of trusted advisers to help them cope with foreign policy. "Kitchen cabinets" of this kind are useful, but they cannot by themselves manage a major international crisis. To the extent that policy calls for extensive diplomatic contacts, military alerts, or actual operations, leaders must of necessity rely on the assistance of the relevant bureaucracies. Political leaders whose demands clash with normal bureaucratic modes of operation are almost certain to encounter resistance.

Even well-run military organizations are likely to be quite rigid. They are avowedly hierarchical, steeped in tradition, and dependent on complex routines. In crisis, moreover, military leaders also tend to emphasize objectives different from those of political leaders. For these reasons, the history of postwar crisis management is filled with instances of civil-military conflict. Here I outline five of the more important organizational sources of such conflict.

First, the military relies on prepackaged routines. Both superpowers have developed weapons and strategies to respond to the kinds of challenges they believe their countries are most likely to face. But these forces, conventional or nuclear, cannot be maintained at full combat readiness, which is too expensive and impractical, and so they are for the most part kept at much lower levels of alert. Procedures have been devised to bring them up to higher levels of readiness at relatively short notice. The ability of the services, individually and collectively, and presumably of their Soviet counterparts to carry out a wide range of options increases as a function of readiness. As a consequence, military organizations will generally insist on moving up to higher alert levels in time of crisis. Even when fully alerted, however, they are likely to resist (or even prove incapable of implementing) options that differ significantly from those they have previously prepared to execute.

Second, the military "factors" problems. Most military operations are composed of distinct routines. These routines are developed by subunits of the services, each of which has a defined area of responsibility within an overall hierarchical structure. Individuals responsible for preparing or implementing military options are likely to consult with subordinate and superior commands, but they may have little or no contact with horizontal commands that are charged with preparing or implementing other aspects of the same operation. Consequently, decisions made by planners almost inevitably reflect a narrow, even parochial view of the problem and may be formulated entirely in ignorance of the political goals that military action is meant to achieve. Those at the apex of the hierarchy, by comparison, are likely to be more sensitive to the political objectives of the government but less knowledgeable about the details of the operation. The obvious danger for crisis management is that no one in the system is likely to know the full range of behavior that any step up the ladder of escalation entails.

Third, options are "staffed" by middle-level officials. Colonels and their civilian equivalents are generally responsive to a set of criteria very different from those that motivate political leaders. When they formulate options, they shape them in terms of standard operating procedures and tailor them to service capabilities and preferences. The resulting plans are likely to maximize traditional military objectives at the expense of precision, flexibility, control, or other values that political leaders will come to consider critical in crisis. As political leaders are almost certain to be ignorant of the details of these options, they will probably remain unaware of the problems they threaten until they occur. By then, it is often too late to do anything about it.

Fourth, the military resists political "interference" in the planning and execution of actions. It generally views political intervention in the planning process as a threat to organizational independence. Resistance will be greatest when political directives clash with service traditions, preferences, and operating procedures. Even more likely to provoke conflict is political supervision of military operations, which violates two sacrosanct principles of military organization: the chain of command, and the autonomy of the local commander. The chain of command, a mechanism for preserving the hierarchical structure of the military, stipulates that orders should proceed step by step down the hierarchy until they reach the officer responsible for their execution. Traditionally the officer on the spot carries out those orders as he sees fit. This discretionary authority is designed to cope with rapidly changing battlefield conditions.

Finally, the military emphasizes military as opposed to political objectives. The military approach to conflict is dominated by the quest for military superiority. Superiority is seen as the essential condition for deterrence and, should that fail, for winning any war. Because they stress capabilities over intentions, military officers tend to be unaware of, or uninterested in, the political constraints and pressure that affect adversarial leaders. They are correspondingly insensitive to the ways in which their striving for military superiority constitutes an important source of tension with adversaries. In keeping with this outlook, the military is likely to advocate crisis policies that work on the other side's capabilities, not on its intentions. This traditional military outlook is a sure recipe for disaster in the nuclear era, but this is something civilian authorities are more likely to recognize. Attempts they make to restrain military efforts to gain or maintain superiority, or in a crisis to subordinate military advantage or preparedness to broader political goals, are almost certain to meet military opposition.

German mobilization plans on the eve of World War I illustrate the pernicious effects of all of the above-mentioned institutional characteristics of the military. The German Army's almost total autonomy enabled it to plan for war in a political vacuum. When the July crisis came, Germany's political leaders were confronted with a military plan that had been formulated solely with reference to narrow organizational criteria and requirements. They discovered its inadequacy only after it was too late—or so the generals said—to do anything about it. The problem will be discussed in more detail in Chapter 4.

The Cuban missile crisis provides an interesting contrast to the

July crisis. Students of that confrontation have documented the extent to which Kennedy's management of that crisis provoked civilian-military conflict. For the most part, they have taken the military to task for its intransigence.[1] On one level these criticisms are valid; military parochialism interfered with the president's efforts to avoid unduly provoking the Soviet Union, and it could have impeded a diplomatic resolution to the crisis. But the criticisms are also naive. Leopards do not change their spots; it is unrealistic to expect the American military to have been receptive to Kennedy and McNamara's effort to orchestrate and direct the blockade from the White House. "Micromanagement," a novel experience for the Pentagon, was certain to provoke anger and opposition.

The civil-military tensions that arose during the missile crisis resulted from the Kennedy administration's efforts to protect itself against the kinds of institutional mishaps that contributed to war in 1914. Kennedy and McNamara were remarkably prescient about the organizational impediments likely to obstruct their crisis strategy. Kennedy later attributed this in part to his just having read Barbara Tuchman's *The Guns of August*, a book that made a big impression on him.[2]

In a sense, American civil-military relations during the Cuban crisis warrant being viewed in a positive light. What is remarkable in retrospect is neither the extent of the conflict nor the number of unforeseen and potentially disastrous incidents that took place. Rather, it is the fact that the conflict was not more serious and its effects more destabilizing. Kennedy and the military both deserve credit in this regard; Kennedy for his insight, and military leaders for the degree to which, when pushed, they departed from routine and improvised procedures that responded to presidential directives and needs.

The Dilemma of Contradictory Objectives

Organizational rigidity constitutes an enduring but partially controllable threat to the efforts of national leaders to manage crises in accord with their political objectives. A potentially more serious problem is the contradiction between the measures necessary to prevent an accidental or unauthorized firing of a nuclear weapon, on the one hand, and those required to guarantee the country's ability to retaliate after being attacked, on the other. This dilemma becomes particularly acute at high levels of strategic alert. There it constitutes the single most serious cause of potential instability.

The previous chapter described the inherent vulnerability of U.S. command and control to nuclear attack and the uncertainty thus created about the nation's ability to retaliate. According to Bruce Blair, author of the most comprehensive study of this problem, measures introduced in the 1960s to protect against accidental and unauthorized use of nuclear weapons make the prospect of retaliation even more uncertain.[3] These safeguards were at least in part a political response to public concerns of the day. They were implemented at a time when policy makers were ignorant of their real strategic consequences. Blair does not call for their repeal; rather, he wants a series of measures designed to reduce the current vulnerability of command and control. In his opinion, this constitutes the nation's strategic Achilles heel.

While Blair and those who share his point of view worry about the prospects for U.S. retaliation, other students of command and control are more concerned about the problem of control. They fear that procedures already instituted to help ensure retaliation significantly raise the risk of war by accident or miscalculation. The most serious danger in a crisis, in their view, is not the mutual temptation to preempt in order to take advantage of the vulnerability of the adversary's command and control. Rather, it is the difficulty that both sides will have in halting escalation once significant military preparations get under way.

The most forceful exponent of this thesis is Paul Bracken. He warns of the possibility of a nuclear Sarajevo, brought about by the interaction of complex institutions that leaders neither understand nor control. "The lesson of World War I," he observes, "was less that war can come about from the actions of obtuse leaders than that a nation's actions in crisis are profoundly influenced by the security institutions built years before the crisis occurs. The process of alerting and mobilizing forces, and of applying those forces, outran the political control apparatus. It even outran the strategies of the states involved."[4]

Bracken is pessimistic about the superpowers' ability to master an acute crisis between them. Contemporary leaders are, he admits, more aware of the dangers of loss of control than were their predecessors in 1914. But they also command more complex and less predictable military organizations: "Even if today's leaders understand the enormously destructive consequences of war, which are far more apparent now than in 1914, the construction of fatalistically complex nuclear command organizations parallels the conflict institutions built in the decade before 1914, but on a far more spectacular and quick-reacting scale."[5]

The command and control systems of the superpowers have matured, as Bracken shows, over the course of the last thirty years. The two most significant developments in this regard have been the vertical and the horizontal integration of these systems. Vertical integration links the warning and intelligence apparatus with the control machinery for nuclear weapons. Horizontal integration consists of tighter central control over nuclear weapons, including the determination of when, if ever, those weapons would be used as well as the targets assigned to them.[6]

Both kinds of integration were prompted by the need to establish greater control over the operational environment. The essential requirement in this regard is timely information about military preparations by adversaries. Such intelligence was made possible by a proliferation of sophisticated electronic and photographic sensors, computerized information processing, complex software algorithms, and control centers.

New procedures also had to be devised to speed up and manage strategic alerts. Thousands of nuclear weapons and their delivery systems, dispersed among numerous scattered commands, necessitated centralized nuclear planning in order to preassign targets and coordinate the delivery of weapons. It required redundant and secure communication channels to keep these forces abreast of adversarial preparations, bring them up to higher states of readiness, or actually authorize them to use their weapons. This requirement made a unified command structure essential.

The result was extraordinarily sensitive and complex systems in which the warning and alerting processes were fused as part of quick-reacting command and control networks. In the United States the institutional heart of this system is the North American Aerospace Defense Command, created in 1957 to serve as the central processor of warning information. NORAD also plays a critical role in an elaborate system of checks and balances designed to prevent accidental or unauthorized use of nuclear weapons. Most important alerting actions or emergency procedures can come into effect only in response to the declaration of the requisite alert level by two separate and independent commands. Institutional checks are built into every level of the system, down to the "two-man rule" that governs the actual launching of a weapon. The strategic bureaucracy deliberately replicates the guiding principle of the federal government, the separation of powers, ensuring that important decisions require the cooperative action of two separate branches.

So far, the combination of institutional checks and physical restraints have prevented an accidental or unauthorized launch. The possibility of such a disaster seems extremely remote under day-to-day operating conditions. It is much more likely to occur in a crisis, however, because of the stresses to which the system would be exposed and the ways in which stress could aggravate structural problems. The system could then behave in unprecedented and dysfunctional ways. We can briefly summarize some of the more troubling and intractable problems of the U.S. alert and response system.

Information Overload. Since the 1960s photographic, electronic, and signals intelligence has been producing a growing stream of data, a stream beyond the ability of existing centers to process. More and more fusion and operations centers were set up to cope with this influx, but they have never really been able to do so. In a crisis, sensory overload could degrade the timeliness and quality of intelligence, increasing the possibility of faulty assessment. Beyond the confines of the relatively streamlined nuclear alerting system, moreover, the problem of overload is even more acute. In 1984 the Defense Department reported that its communication networks had transmitted 56.7 million messages, and this total did not include voice transmissions or information in data-processing systems. In time of crisis, standing orders prohibit the transmission of nonessential administrative messages. Even so, operational officers worry, it could take five to ten hours for critical messages to get through.[7]

Institutional Compartmentalization. Fusion and control centers have proliferated, a growth paralleled by an enormous expansion of the strategic forces and their command centers. The result has been a large and highly fragmented strategic bureaucracy. It is now impossible for any single individual to develop intimate knowledge of the entire system and its component parts. The view of the people within the system has accordingly become increasingly compartmentalized and parochial. Coordination becomes more difficult, heightening the prospect that in a novel situation, components of the system would work at cross-purposes or in ways that would produce entirely unanticipated consequences.

Compression of Decision Time. In the 1950s the United States expected to have many hours of advance warning of nuclear attack. Warning today is measured in minutes; missiles fired from offshore submarines would take only five to eight minutes to destroy many of the nation's most important command and control centers, submarine pens, and bomber bases. A variety of systems has been developed to provide more timely warning of attack. Probably the most important of these is a constellation of satellites with infrared sensors capable of detecting Soviet ICBM and SLBM launches. These satellites transmit real-time data to ground stations in Colorado, Germany, and Australia which in turn send it not only to NORAD but also to SAC and the National Military Command Center in the Pentagon. Presumably all three centers have the authority to act on the basis of a warning conveyed in this manner. Time pressure has necessitated some circumvention of institutional checks and balances. In the future it may compel even more far-reaching breaches of these organizational safeguards.

Informal Procedures. Any complex system subject to severe time constraints has difficulty in performing its assigned task if it is run strictly by the book. In practice, informal understandings develop among its operators. These take the form of alternative and simplified procedures that facilitate performance by circumventing cumbersome institutional safeguards. The military, Bracken reports, has resorted to such shortcuts; officers have discovered ways of getting around many of the formal requirements of the nuclear command organization in order to carry out their assigned missions. Alternative procedures of this kind, established by oral agreement, are invisible from the outside until they are put into practice. So we do not know how widespread they have become; but we can be sure they have undermined some of the most important safeguards built into the system.

A similar development within the Soviet Union has paralleled the horizontal and vertical integration of the U.S. strategic warning and response system. The two systems have also become tightly coupled. This link was initially a side effect of the capability of each superpower to track the other's military preparations. But now both superpowers also have a growing capability to monitor each other's warning systems. As the two alert and response systems interact with each other on something close to an instantaneous basis, they can be described as components of a single gigantic nuclear system.

On a day-to-day basis the tight coupling of the two alert and response systems is not destabilizing. Each superpower can observe—that is, verify—that the other's forces are operating in the normal matter. Soviet forces are also kept at a lower state of readiness than American forces, something that also minimizes interaction between the two systems. Most important of all, both systems, it is believed, function in a dampening mode; they work on the supposition that something is wrong with information indicating an enemy attack. In the case of the United States the system about which we know more, a warning triggers actions designed to test and (it is hoped) discredit it before it is accepted as the basis for action. Finally, all the organizational and physical safeguards that both sides have instituted to prevent an accidental or unauthorized launch are fully in effect.

For all of these reasons, the United States had coped successfully with over fifteen hundred false alarms. These have been triggered by causes as diverse as a misleading radar image, the erroneous broadcast of a training tape indicating that a nuclear attack was under way, and the failure of a computer chip which prompted transmission of the same message. None of these events triggered a war because of what the military, with its love of jargon, calls "dual phenomenology": the reliance on two or more independent sensors to confirm an attack. Nevertheless, many of these incidents triggered some preparatory measures before they were diagnosed as false alerts. In 1980, in response to the computer chip failure, NEACP and B-52s were readied for take-off, and the airborne command post of the U.S. commander in the Pacific was actually launched.[8]

Incidents such as these illustrate the propensity of tightly coupled systems to produce overcompensating effects.[9] Bracken reports that additional alerting and preparatory measures would have gone into effect had the computer chip error been discovered a little more belatedly. Almost a hundred B-52s would have been launched and sent north toward their failsafe position, and alert messages would have been sent to ICBM crews as well as to conventional forces in Europe and Korea. The president would have been awakened at approximately 2:30 A.M. and told that he had minutes to evacuate Washington, decide on the country's war plan, and order a response.[10] Even these actions would likely not have led to war. They could have triggered Soviet counterpreparations, however, which would have brought both superpowers to unnecessary and dangerously high levels of nuclear readiness.

The more serious threat is posed by an accident or miscalculation in time of crisis. During a period of high tension more sensors would be activated and assigned to strategic warning. The stream of data entering superpower command and control channels would increase manyfold, and the two warning and response systems would become coupled even more tightly. Military and civilian leaders would also be more disposed to give credence to warnings of attack. If the crisis became sufficiently acute, it might in effect reverse the procedure for responding to warnings: policy makers might authorize a significant level of response *before* ascertaining the validity of a warning. A Soviet response could then provide an American counterresponse. . . . A false alarm could accordingly prompt the superpowers to move up the strategic escalation ladder to dangerously high levels of alert.

Bracken breaks new ground in our understanding of the dangers inherent in superpower crises. Together with the pioneering work of John Steinbruner, his book ought to alert policy makers and military planners to problems that until recently they have largely ignored.[11] We can help by specifying the different ways by which system malfunctions could lead to war, which would impart verisimilitude to the somewhat abstract warnings that Steinbruner and Bracken provide. It would also be useful in a practical sense, because different kinds of malfunction almost certainly lead to war in different ways. By identifying these different paths to war, we would take the first step toward making some judgments about their probability and most likely causes.

Such a task is clearly beyond the scope of this book. I do intend to make a start in this direction, however, by describing three generic ways in which a system malfunction could provoke a war. These are mistaken retaliation, war by chain reaction, and war precipitated by third parties. I also examine possible scenarios for each of these paths to war. These scenarios should help us think more clearly about the problem and possible means for coping with it.

Mistaken Retaliation

The first possibility to consider arises from the near certainty that in the course of an acute crisis, the president would choose to predelegate some degree of launch authority. At least the theoretical possibility would thereby exist that some officer could use or order the use of nuclear weapons in the absence of a command to do so by the National Command Authority (NCA).

To dramatize the implications of predelegation, Bracken invokes the metaphor of a revolver. A revolver has two control mechanisms: a safety catch and a trigger. As long as the safety catch is locked, the trigger cannot fire the gun. Once the catch is released, control of the weapon passes to the trigger. The strategic arsenal can be likened to a revolver with one safety catch and many triggers. When negative control is in effect, the safety catch is engaged; none of the triggers can fire a nuclear weapon. However, the vulnerability of the NCA to attack requires that arrangements be made to predelegate launch authority, either in advance of an attack or in response to indications that one has started. Presumably, the more acute the crisis and the perceptions of vulnerability, the further down the chain of command would launch authority be predelegated. If and when the strategic system is shifted to positive control, any one of these triggers could fire the nuclear gun.

Very little is known in practice about delegation of launch authority. The subject is highly sensitive, and the few officials, past and present, who have any knowledge of it have been understandably reluctant to speak out. Congressional inquiries have established that the president alone has the authority to order the use of nuclear weapons, but that he may delegate this power without limit.[12] Launch authority is no different constitutionally from any of the other executive powers of the president.

Successive administrations have refused to comment on predelegation, but some information has nevertheless emerged from the statements of former officials. General Earl E. Partridge, a former commander of NORAD, admitted in a 1957 interview that he had been given the power to use nuclear weapons in specified emergency situations.[13] General Lauris Norstad, former NATO commander, hinted to journalists that he had been granted similar leeway by President Kennedy.[14] Daniel Ellsberg insists that presidents Eisenhower, Kennedy, and Johnson delegated launch authority to six or seven 3- and 4-star generals, the officers in charge of each of the unified and specified nuclear commands. He claims to have uncovered presidential letters authorizing these officers to use these weapons on their own authority in the aftermath of a Soviet attack.[15] Evidence unearthed by congressional hearings in 1975 and 1976 seems to bear out Ellsberg's claim.[16] Looking Glass, usually commanded by a 2-star SAC general, may also be able to initiate a launch order. Former National Security Agency official Raymond Tate has admitted that the airborne command posts and Looking Glass possess the authorization codes. If so, they can not only verify and relay NCA commands but also take action on their own.[17]

After reviewing all the available public evidence, Paul Bracken concludes that predelegation of launch authority extends at least as far down the chain of command as the unified and specified commands. In time of crisis, when forces are being brought to a high state of readiness, he believes, predelegation would almost certainly have to extend further down the chain of command. In Europe the extreme vulnerability of command centers to almost instantaneous destruction appears to necessitate more widely distributed launch authority in order to guarantee retaliation.[18]

No delegation of launch authority would be entirely discretionary; presidents are intensely jealous of their prerogatives and do not willingly relinquish authority. All kinds of restrictions would almost certainly hedge in any prior authorization to use nuclear weapons. They would be designed to preserve presidential authority as well as guard against the ill-considered use of such weapons. But whatever the restrictions imposed, the very fact of predelegation increases the risk that missiles might be launched.

In a serious crisis, and even more so in a conventional war, numerous requests for release authority would work their way up the chain of command. Some students of the alerting process contend that these requests would start to come in earlier than is generally expected. Commanders would want to have authorizations in their pocket for fear it would take too long to obtain them when and if they actually needed them.[19] The NCA would process requests for release authority and, depending upon the circumstances, could grant some or all of them. It is even conceivable that the president, overburdened by political responsibilities, would allow the secretary of defense to pass on the merits of the requests.

Grants of release authority could lead to hundreds or even thousands of armed weapons in the hands of field-grade officers. Pershing II or nuclear artillery batteries that were under attack and in danger of being overrun could launch their weapons, perhaps in the erroneous belief that a nuclear war had already begun. One weapon fired in this manner might be enough by itself to start a full-scale nuclear war. Alternatively, it could set in motion a chain of reprisals leading quickly to full-scale theater nuclear war.

Such a chain of events could bring nuclear war even in the absence of conventional conflict. The most likely initiator would be a submarine. American sea-launched ballistic missiles lack the physical locks (PALs) that protect other nuclear weapons. To prevent an unauthorized launch, the navy relies instead on institutional safeguards. The launching of a missile requires the concerted action of four officers and eleven seamen at various stations of the boat. All four officers must throw switches or turn keys within moments of one another; the missile cannot be fired if any one of them fails or refuses to participate.[20]

The navy insists that an SSBN could launch a missile only in response to an Emergency Action Message sent by the NCA. In practice, however, the navy recognizes the difficulty of communicating with submarines on station under wartime conditions and the vulnerability of all existing means for doing so. No doubt this is why naval officials have consistently opposed the introduction of PALs on submarines; such an innovation would necessitate a coded message from the NCA to arm the missiles. The navy's position on the matter encourages speculation that submarine captains must have secret standing orders to execute nuclear operations, in certain well-defined conditions, without an EAM. This supposition is supported by what little evidence exists in the public record.[21]

The real restraint on an SSBN captain is the crew; the captain would have to convince *them* that a launch was warranted. They might agree if they believed that nuclear war had already begun and that the NCA had been destroyed. A communications failure could erroneously engender such a conclusion, especially at the height of an acute crisis. The implications to the crew would be all the more alarming if the submarine had been attacked or had detected a Soviet effort to track and acquire it as a target. This is no mere speculation; the United States forced Soviet submarines to surface during the Cuban missile crisis and was prepared to use depth charges to do so. Conceivably the Soviets could do the same in some future confrontation if war seemed imminent or even highly likely.

An SSBN being hunted or trailed by the Soviets would confront a dilemma. The ambiguity of the situation would dictate careful efforts to establish what was really going on, but those efforts would require the submarine to come close to the surface, in order to trail or put up an antenna. Either action would make the boat vulnerable, the very last thing its captain would want to do in the circumstances. A safer course of action would be to try to evade Soviet units in the vicinity and only afterward attempt to communicate with other elements of the fleet or the NCA. If for any reason the submarine was unable to break away or communicate with the outside, its captain would find himself in a delicate position. He would be sorely tempted to launch his missiles if he believed his submarine was about to be destroyed. One Ohio-class submarine carries twenty-four missiles, each with four warheads, some of them almost certainly aimed at Soviet cities or targets in or near cities. Devastation of Soviet cities would be more than ample provocation to trigger a major nuclear reprisal against the United States. It is possible that misleading circumstantial evidence could lead a submarine to start a nuclear war when its captain and crew believed that they were retaliating.

There is a second submarine scenario to consider. Unlike the United States, the Soviet Union keeps most of its SSBNs in home waters. The boats may constitute a reserve strategic force; certainly the Soviets have made elaborate efforts to develop the naval and air forces necessary to protect them.[22] The American Navy has bridled at the notion of a Soviet SSBN sanctuary, and since John Lehman became secretary of the navy at the onset of the Reagan administration, it has seemingly become committed to going after the Soviet Navy in its home waters. For American attack submarines, this has long been routine practice.[23] It has been publicly reported that they routinely patrol in areas where Soviet SSBNs would seek to hide. They are also known to have penetrated Soviet territorial waters in the vicinity of major Soviet naval bases and on several occasions to have collided with Soviet vessels.[24]

Aggressive patrolling of this kind, in the opinion of retired admiral Worth Bagley, "risks an incident in time of tension."[25] The Soviets could interpret it as an indication that the United States thought war was likely and was putting itself in a more advantageous position to fight it. If American attack submarines appeared to threaten its SSBNs, the Soviet Navy would be chaffing at the bit to force them to back off. A Soviet move of this kind would, under the tensest of situations, risk a naval encounter. An engagement between the Soviet Navy and American submarines could easily involve nuclear weapons. The U.S. rules of engagement apparently permit submarine captains to protect themselves *in extremis* with nuclear weapons.[26]

War by Chain Reaction

The second way in which war could arise from system malfunction is as the result of an escalating spiral that leads to war by accident or preemption. The previous scenarios assumed that strategic forces were already at a high level of readiness; this sequence, by contrast, is itself the cause of such an alert.

High levels of strategic alert can result from an action-reaction cycle between Soviets and Americans which drives both superpowers to higher levels of military readiness than either independently would have chosen to implement. As Paul Bracken notes,

> a threatening Soviet military action or alert can be detected almost immediately by American warning and intelligence systems and conveyed to force commanders. The detected action may not have a clear meaning, but because of its possible consequences protective measures must be taken against it. The action-reaction process does not necessarily stop after only two moves, however. It can proceed to many moves and can, and often does, extend to air- and land-based forces because of the effect of tight coupling. In certain political and military situations, this action-reaction process can be described as a cat-and-mouse game of maneuvering for geographic and tactical position. In more ominous circumstances, it may be seen as a jockeying for positions before the first salvo of an all-out war.[27]

Such a phenomenon is made more probable by the growing capability of the American and Soviet alerting systems to monitor each other, not just the military actions that both systems produce. The effect is to compress the action-reaction cycle. It may also intensify it, because the inherent ambiguity of intelligence of this kind permits, or even encourages, the most threatening kinds of interpretation. High alert levels, regardless of how they come about, would seriously interfere with whatever efforts were under way to resolve the crisis. To quote Bracken again:

> The interdependencies and synergies that were safely ignored during the peacetime cat-and-mouse game then begin to enter the picture. Tight coupling of the forces increases, information begins to inundate headquarters, and human, preprogrammed-computers, and organizational responses are invoked. Each response, whether it arises from a human operator or a computer, is intended to meet some narrow precautionary objective, but the overall effect of both Soviet and American actions might be to aggravate the crisis, forcing alert levels to ratchet upward worldwide. Although each side might well believe it was taking necessary precautionary moves, the other side might see a precaution as a threat. This in turn clicks the alert level upward another notch.[28]

High levels of readiness threaten to spill over into war in two distinct ways. By definition, they entail some degree of strategic force readiness, dispersal of theater nuclear systems, activation of wartime command posts, and possibly even the spontaneous or organized evacuation of cities. The other side could readily interpret actions of this kind as indications of an impending attack. Such actions could prompt preemption, for any of the reasons analyzed in the previous chapter.

Strategic alerts also risk war by mistaken retaliation. Either side could initiate hostilities in the mistaken belief that it was responding to an enemy attack. We have already described two scenarios of this kind, both of them the result of system failures or misjudgment at the bottom of the nuclear chain of command. The same thing could also occur at the national command level, taking the form of a launch in response to a false warning of attack.

The vertical integration of the intelligence systems responsible for providing warning of attack permits the linking together or "fusing" of data collected from diverse sources into a more comprehensive assessment of threat. The ability to do so increases in a crisis, because warning of attack is given a much higher intelligence priority. Because of well-documented human propensity to impose order on events and to see in them a meaning deeper than reality often warrants, moreover, analysts will tend to view the resulting information flow as pieces of a puzzle.[29] An exaggerated threat assessment could result. Events that in ordinary circumstances might be seen as unrelated would more likely be interpreted as part of some coordinated plan by the adversary.[30]

Physical or human failures in the alerting system, in tandem with an exaggerated perception of threat, could prompt extreme overreaction. Suppose, for example, that BMEWS or one of the PAVE PAWS radars is down because of scheduled maintenance or a failure of critical components. Frantic efforts to bring it back on line began at the onset of the crisis, but they have been unsuccessful. As the political situation begins to unravel, a U.S. early warning satellite reports that an SSBN, several hundred miles off the east coast, has just launched seven missiles. (Erroneous reports of this kind have sometimes been received during severe thunderstorms.)

With the relevant PAVE PAWS down and the missiles out of range of the older FSS-7 radars, NORAD would be in a quandary. Most of the other available radars would be useless, because the small MIRV payloads of modern SLBMs are invisible to them. To wait for PARCS in Grand Forks to verify that the United States was really under attack would waste precious minutes in which Soviet warheads could be zeroing in on their targets. To be on the safe side, NORAD and SAC might just arm the Minutemen, send the bombers aloft, and order additional communications aircraft aloft. They would probably also convene a "missile attack conference" with the president.

The Soviets possess the capability to monitor our alert levels and would detect some or all of the U.S. responses to the satellite warning. Not knowing that these were precautionary measures initiated in response to a false alert, Soviet officials might fear they were offensive in intent. The Soviets might move to a higher state of strategic readiness. Some analysts contend that U.S. intelligence is good enough to monitor Soviet strategic communications and to report that the Strategic Rocket Force, and perhaps theater nuclear forces as well, have received some kind of alert order or action message. This information, coming hard on the heels of the still unconfirmed satellite report, would intensify U.S. fears that

SLBMs had actually been launched—that they were a precursor to a massive ICBM assault. After all, we would have no way of knowing that the Soviet action was only a response to our alert. An anxious president could order further steps to improve the prospects of retaliation, bringing the country to the very brink of war.

Up to this point our imaginary scenario has involved no human error or unanticipated interaction of components of the warning system. The United States and the Soviet Union have progressed toward full wartime readiness as the result of deliberate decisions taken in response to one faulty sensor. But if, in addition to time constraints and tight coupling, we allow for some kind of system malfunction, compounded perhaps by human error, the chance of catastrophic failure increases enormously.

What, for example, if the training tape that caused the famous 1979 alert had been fed into the computer during a crisis? At that time the monitors at NORAD, the National Military Command Center in the Pentagon, the Pacific Command, and elsewhere showed that the United States was under attack from both land- and sea-based missiles. Such a tape would have only confirmed a misleading satellite report. In 1979 it took six minutes to determine what was wrong with the system, although only about a minute, the air force alleges, to realize that something had gone awry. At the height of a crisis, however, even that one minute would be more than enough time for a president mistakenly to contemplate retaliation against the Soviet Union.

SAC officers contend that it could not happen, because an exercise would never be conducted during a crisis. Perhaps they are right. But if asked beforehand, they doubtless would have sworn that the computer on which the tape was played could never have fed into the on-line warning system—something that inexplicably *did* happen. My point is not that the event will repeat itself, but rather that there is tremendous uncertainty about how such a complex and tightly coupled system would perform in a crisis. A faulty computer chip, poorly designed software, an operator error, any one of a hundred things or, worse still, a combination of them could result in an accidental war.

Representatives of NORAD dismiss these concerns and the scenarios associated with them as extremely unrealistic. They stress the low probability of component failures and the redundancy built in to all critical operations of the system. In their view a system failure has a minuscule probability of occurring, because it would have to be the multiple of a large number of individual failures. Even if they are right—and there is good reason to doubt it—they can give no assurance that a catastrophic failure could not happen.

Nuclear Alert and Response: A High-Risk System

The history of complex systems is punctuated by disquieting examples of failures that the designers and operators of the system were convinced could never happen. Most of them, including the Great Northeast Blackout of November 1965 and the Three Mile Island incident of 28 March 1979, resulted from the synergistic interaction of physical and human errors. The effects of initial component or subsystem failures were compounded by the failure of system operators to recognize the nature and extent of the problems because the failures were entirely unanticipated. In the case of Three Mile Island and the DC-10 crash of 25 May 1979 at O'Hare airport in Chicago, which was caused by the loss of an engine, human intervention based on faulty diagnosis of the problem greatly aggravated the effects of the failures.[31]

High-risk systems have characteristics that make them especially dangerous. Charles Perrow, author of a highly regarded comparison of high-risk technologies, argues that the danger stems from the way in which failures can interact and magnify their consequences. According to Perrow,

> The argument is basically very simple. We start with a plant, airplane, ship, biology laboratory, or other setting with a lot of components (parts, procedures, operators). Then we need two or more failures among components that interact in some unexpected way. No one dreamed that when X failed, Y would also be out of order and the two failures would interact so as to both start a fire and silence the fire alarm. Furthermore, no one can figure out the interaction at the time and thus know what to do. The problem is just something that never occurred to the designers. Next time they will put in an extra alarm system and a fire suppressor, but who knows, that might just allow three more unexpected interactions among inevitable failures. This interacting tendency is a characteristic of a system, not of a part or an operator.[32]

In some systems, complexity of this kind is not necessarily fatal: accidents do not threaten human life, or their effects can somehow be minimized. There is enough available slack, spare time, and other ways of getting things done. But in tightly coupled systems, processes take place very quickly. They cannot easily be halted, and it is correspondingly difficult to isolate their effects.

Better equipment or different procedures can sometimes reduce the tightness of the coupling or minimize its effects. In air traffic control, for example, better organization and equipment have reduced interactive complexity and tight coupling in recent years. In other systems, among them nuclear power plants, neither technical improvements nor organizational innovations have succeeded in making them less accident-prone. The reason, according to Perrow, is that such systems are based on organizational structures that have "large internal contradictions." Procedural changes cannot resolve these contradictions. Technological "improvements" only increase interactive complexity and tighten the coupling, making the system even more susceptible to certain kinds of accidents.[33]

Nuclear response systems have several characteristics that make them appear even more accident-prone than many other kinds of high-risk systems. First, we are dealing with *two* systems, not just one. Individually, the strategic alerting networks of the superpowers have been described as the most complex systems ever designed by humans. When those two networks are considered together, they constitute another quantum leap in complexity.

Second, a catastrophic failure, if it occurred, could result not from one or the other national alert and response system, but from the interaction of the two. But we know next to nothing about how these systems would interact. What limited knowledge we do have is based on their performance at the lowest levels of alert, which is not very useful in predicting interactions at times of crisis. In the absence of detailed information about Soviet alert and response procedures, simulation will not help us either. Even if this information were available, simulations would still be of only limited utility in "debugging" the system, because they cannot begin to replicate anything close to the full range of possible interactions. Nor could they reliably factor in Soviet decision-making patterns and perspectives, which, by all accounts, are quite different from ours.

Third, evidence from the purely national level gives us little reason for confidence in the reliability of either side's warning system. The American system, as we have seen, has experienced numerous difficulties, some of them attributable to violations of procedure, others to the failure or malfunction of physical components. Many of these problems, according to General William Hilsman, former head of the Defense Communications Agency, are the kinds of things that "everybody would have told you were technically impossible."[34] Critics of the system have also pointed to organizational and technical shortcomings; among them the computer software has come in for special criticism.[35] Unreliable software is a particularly serious matter, because the nature of any problem might not become apparent until after something had gone seriously wrong.[36]

The Soviet early warning system is less redundant and less technically sophisticated than its American counterpart. Up to 1986 the Soviets have never been able to maintain a full constellation of early warning satellites in orbit.[37] Their over-the-horizon radars are not connected to anything close to state-of-the-art technology for processing information. Nor do they provide coverage of all of the azimuths from which the Soviet Union can be attacked.[38] Their sensors have a propensity for malfunction even greater than that of U.S. sensors. The Soviets also operate under greater time constraints, because a large percentage of their retaliatory forces consists of vulnerable land-based ICBMs.

Complex Soviet systems are, there is reason to suspect, simply less reliable than their American counterparts. Soviet air transport is notoriously unsafe. Soviet airliners lack many of the safety features of American-built craft and generally do not have the same level of redundancy in vital subsystems. Soviet air control systems and procedures are also reported to be much less sophisticated. Soviet nuclear plants were known to be another problem area even before the meltdown at Chernobyl in April 1986. Many incorporate even fewer safety devices than Chernobyl and most American plants. In 1957–58 a major nuclear accident in Kyshtym in the Urals apparently contaminated a wide area with radiation. A Soviet biological and chemical weapons facility in Novosibirsk is reported to have had a major accident in April–May 1978 leading to the death of three hundred people.[39] Soviet missiles also have a bad safety record.[40]

To their credit, Soviet political leaders appear to take at least as great an interest in the internal control of nuclear weapons as do their American counterparts. Deep distrust of the loyalty of subordinates is one of the hallmarks of the Soviet system, and it has prompted Soviet leaders to maximize their control at every level of the formulation and implementation of policy. American leaders have always felt reasonably confident about their ability to prevent an unauthorized launch; Soviet leaders apparently do not. This lack of confidence may help to explain why the Soviets have continued to rely so heavily on land-based missiles; such assets are more vulnerable than submarines, but they are also easier to control from Moscow. It may also explain why Soviet forces were traditionally kept at a lower state of readiness than American forces.[41]

There is further evidence of Soviet concern with maintaining tight central control over nuclear weapons. Consider, for example, close involvement of the Committee on State Security (KGB) in their development and deployment. The KGB managed the Soviet nuclear weapons program until after Stalin's death; at least one Soviet source maintains that nuclear weapons did not become available to the military until 1954.[42] Even then the KGB controlled and guarded nuclear bombs and warheads at depots that were often well removed from weapons sites. As for missile systems deployed in Eastern Europe, the warheads and bombs that they would carry are retained, it is reported, inside the Soviet Union. Not until the mid-1960s did the Soviets place nuclear warheads on ICBMs on day-to-day alert.[43]

Third- and fourth-generation Soviet missiles and their control centers allegedly have electronic safeguards akin to PALs.[44] The KGB nevertheless continues to play an important role in protecting against unauthorized use of nuclear weapons. KGB troops still maintain control over storage depots for these weapons, and they probably have instructions not to hand them over to the military until verifying orders through their own, independent communication channels to Soviet political leaders.[45] Reports suggest, furthermore, that Soviet ICBM launch centers have four-man crews—two of them KGB officers.[46]

This Soviet emphasis on control makes it likely that Soviet delegation of launch authority, if it takes place at all, will occur at a later stage in a crisis than the equivalent American decision.[47] It can be considered a plus for crisis stability. When Soviet devolution occurs, however, or once warheads are married up to theater-based delivery systems, control is likely to become precarious very quickly. Stephen Meyer ventures the judgment that because of its concern for control, "the Soviet political leadership would withhold nuclear release authority until prepared to see the military actually use weapons."[48]

In the United States the transition from negative to positive control is envisaged by defense officials as a smooth if anxiety-provoking transition. Soviet leaders, in constrast, appear to believe that as they move toward war readiness, at some point they will experience a sharp discontinuity in their ability to maintain control over Soviet nuclear forces. This difference in attitude holds implications for crisis stability which cut two ways. Soviet determination to maintain central control as long as possible discourages high alert levels, and to that extent it is beneficial. But it is clearly destabilizing once these levels are reached—and all the more so if American leaders are aware of the problem. It gives both sides a strong incentive to preempt; the Soviets out of fear that it is their last clear chance to mount a coordinated attack, and the Americans out of fear that the Soviets are about to do so.

A fourth factor that makes alert and response systems appear accident-prone is that neither superpower has ever had a full dress rehearsal. We do not know how they would perform in a real crisis. To be sure, parts of the American system, and no doubt the Soviet system as well, have been tested in exercises and simulations. Exercises almost always reveal shortcomings the first time any reasonably complicated procedure goes into effect. Many other kinds of complex systems can be tested repeatedly, however, while obvious political constraints prevent alert and response systems being frequently tested. Even infrequent rehearsals of all or significant parts of the American system would arouse serious political opposition. It could also convey the impression to West European allies and the Soviet Union that the United States was turning into a militarized society seriously contemplating nuclear war. Rehearsals could themselves be destabilizing; they would likely prompt a range of precautionary alerting measures by the other side.

These obstacles put a premium on limited tests and simulations as means of debugging the alert and warning systems. Unfortunately, such methods may not be being exploited to their fullest potential. Some retired officers who have participated have confided that American tests and simulations are seldom based on realistic assumptions. One reason for this, they allege, is concern that failure of any components under operational conditions would expose NORAD to congressional criticism and, as a result, to closer scrutiny.

The fifth and final point to ponder is the implications of the reluctance of the operators to admit to others—and perhaps to themselves—the full extent of the system's vulnerabilities. This is disturbingly reminiscent of pre–World War I Germany, where war games were consistently rigged in order to demonstrate the feasibility of the Schlieffen Plan. The general staff subsequently used the results of these exercises to convince itself of the wisdom of its strategy.[49] Everything went well until the plan had to be executed against an adversary who was not predisposed to conspire with the general staff to bring about the plan's success.

The nuclear navy and SAC are bastions of "technological optimists."[50] Officers in both believe that their weapons and their command and control are reliable and robust. Most of them are reluctant, even unwilling, to consider the possibility of a catastrophic failure, especially one triggered by human error. Their attitude is attributable in the first instance to cognitive bias; these officers work with these systems every day and have developed a high degree of confidence in them. But the bias is also motivated, because the recognition that war could arise from component failure, operator error, or disastrously faulty judgment by the NCA must be unsettling in the highest degree.

Critics, by contrast, are invariably "systems pessimists." They have adopted this perspective as the result of *their* experience, usually personal or historical knowledge of complex systems that have failed. Critics tend to argue by analogy; they draw parallels between the system under discussion and ones that have failed. System operators generally find this kind of argument unconvincing; they are prepared to listen only to scenarios that describe ways their specific system could fail and the reasons for it. The Nuclear Regulatory Commission, for example, refused to take industry critics seriously until they pinpointed the possibility of specific component failures and consequences that could lead to wider system failure. Even then, most of their dire predictions were dismissed as utterly fanciful. The critics began to gain credibility only after Three Mile Island and other less dramatic breakdowns of the kind the industry said could never happen.[51]

In the case of strategic warning, critics can never become as conversant with the system as its operators. So many important details are highly classified, and will remain so, that outsiders inevitably lack credibility in the eyes of the operators. This is alarming. The misplaced faith of the operators of the warning and response system in its inherent robustness could contribute significantly to some future system failure.

War Precipitated by a Third Party

In 1914 third parties were instrumental both in bringing about the crisis and also in preventing its resolution. A similar threat exists today. Tensions between the superpowers could be aggravated dramatically by either the independent military actions of a client state or a nuclear alert by a third power. The 1973 Middle East war testifies to the dangers of the former case. The Israelis succeeded in cutting off and almost completely surrounding the two Egyptian armies opposing them, touching off near panic in Cairo and deep concern in Moscow. The Soviet threat to intervene, which touched off the most serious superpower crisis since 1962, was designed to stop the Israeli advance by putting pressure on the United States to restrain its client. An effective cease-fire put an end to the crisis and left moot the question of what Moscow would have done had Israel resisted American appeals to halt its offensive on the Egyptian front.[52]

The other danger that third parties pose stems from the proliferation of nuclear weapons. In a superpower crisis, especially one focused on Europe, the British and French nuclear arsenals must also be taken into account. In the Far East there is China to contend with. Very little is publicly known about the alert procedures of any of these powers or about the circumstances in which they would bring their strategic forces up to wartime readiness. Nor do we know how the Soviet Union would respond to force generation by any of them.

Britain is, from the perspective of crisis stability, probably the least worrisome of the three countries. Its strategic forces consist of aircraft- and submarine-based missiles. Submarines are under relatively little pressure to launch immediately because their commanders expect them to survive the initial round of a nuclear war. However, their communication links to national command authorities are extremely vulnerable, giving rise to the same kind of problems we encountered in the case of U.S. submarine force. This vulnerability could generate pressures for preemption. There is also the danger of mistaken retaliation. Britain is replacing its aging fleet of Poseidons with Ohio-class boats, each of which will carry twenty-four MIRVed missiles. As most of these warheads will continue to be targeted against Soviet cities, just one mishap probably will prompt full-scale reprisal. Soviet leaders, if they were still alive, would not necessarily know that it was a British rather than an American SSBN which had attacked them. The United States could end up paying a horrendous price for an allied mistake, which makes it imperative that we learn more about the British command and control system and its susceptibilities to malfunction.[53]

France poses a different kind of problem. The *force de dissuasion* includes nuclear-capable aircraft as well as submarine- and land-based missiles. The missiles are deployed in underground silos on the Plateau d'Albion where they are vulnerable to attack from Soviet theater and intercontinental missiles. They are scheduled to be replaced by less vulnerable, mobile missiles sometime in the 1990s. The short flight time of the Soviet SS-20, and the even shorter flight time of follow-on missile systems that could be launched from East Germany or Czechoslovakia, puts pressure on the French to develop some option along the lines of launch on warning or launch under attack. A hair trigger of this kind is highly destabilizing for all the reasons discussed in the previous chapter. Crisis stability would be better served by a different strategy: ride out an attack, rely instead on the retaliatory power of the more survivable submarine force to deter the Soviet Union from attacking the land-based missile force.[54]

British forces are integrated into NATO but can still be used independently, as the war in the Falklands demonstrated. It is difficult, however, to conceive of a scenario in which the British government would contemplate unilateral use of nuclear weapons against the Soviet Union. France's forces are independent of NATO but presumably would not be used unless the territorial integrity of France were directly threatened.[55] By then, NATO might already have made a decision to go nuclear. The more important consideration for our purposes is the nature of French alerting procedures and the circumstances in which the French would bring nuclear forces, strategic and tactical, up to higher states of readiness. If France were to go to a higher state of alert than the United States or alerted its forces earlier, it could make a Soviet-American crisis centered in Europe much more difficult to resolve.

Even less is known about Chinese nuclear policies and alert procedures, which constitute another possible source of instability.[56] China is not an American ally, nor are Chinese and American security policies coordinated. A Chinese nuclear alert would still prompt some kind of Soviet response, and certain circumstances might even tempt the Soviets to launch a preemptive attack. A Soviet strike against the Chinese, even just a counteralert, would prompt the United States to raise its own state of military readiness. Extreme risk lies in the possibility of setting in motion a spiral of escalating interactions between the alert and response systems of the two superpowers.

Soviet and American strategic forces thus remain vulnerable not only to each other but also to the forces and policies of third parties. So many uncertainties surround the consequences of the interaction of superpower strategic forces that we cannot with confidence render any kind of judgment about the risks that various levels of alert entail. But we can be certain that the existence of third party nuclear forces makes those risks greater than they would otherwise be.

Conclusions

Crises, this chapter has shown, could lead to unintended nuclear war in several different ways. Most disturbing of all in this regard is the link they reveal between preemption and loss of control. For the most important means of protecting against the one makes the other more likely.

Because there are no observable differences between military preparations for defensive purposes and those for offensive purposes, both superpowers, to be on the safe side, generally assume the worst about adversarial intentions. Such assumptions could easily lead them to carry out military preparations in response to any strategic alert by their adversary. By doing so, they reduce their vulnerability to preemption and thereby their adversary's incentive for striking first. However, high levels of alert risk accidental war because of all the difficulties involved in controlling alerted forces in a crisis. Thus military preparations that began as a purely defensive response to a threatening situation could provoke a cycle of reaction and counterreaction that could end up triggering an unintended nuclear war.

As if this were not bad enough, there is the complicating factor of third parties. The United States has only one nuclear adversary: the Soviet Union. By contrast, the Soviet Union is targeted with nuclear weapons by China, Britain, France, and possibly Israel as well. A crisis in the Far East, the Middle East, or Europe could therefore entail nuclear alerts by other countries in addition to the superpowers. Such alerts could provoke a superpower alert. The greater vulnerability of third-party nuclear forces, and their command and control, constitutes another important source of crisis instability. Finally, these forces are likely to be even more accident-prone when alerted because they lack many of the safeguards or redundant communication links that characterize superpower command and control.

The policy lesson of this chapter is clear: it is imperative that the superpowers do their best to refrain from nuclear alerts. But the American defense establishment still conceives of nuclear alerts as an effective means of demonstrating resolve. This was the "lesson" they learned from the 1962 and 1973 crises. Widespread ignorance of the dangers of strategic alerts thus constitutes another important source of instability. It is to this problem that we now turn.

[4]

Miscalculated Escalation

Miscalculated escalation is our third sequence to war. It refers to steps taken up the political-military escalation ladder in a crisis, steps taken to moderate adversarial behavior which instead provoke further escalation. Miscalculated escalation may be the most important sequence to war, because it can be responsible in the first place for a crisis and subsequently for high levels of escalation which threaten war by preemption or loss of control.

I begin with an analysis of different kinds of thresholds and explore their relationship to miscalculated escalation. I then demonstrate how miscalculated escalation was an important cause of war in 1914. This case study reveals two fundamental causes of miscalculated escalation: the inability of policy makers to empathize with their adversaries, and the ignorance of policy makers of their own country's war plans and the plans' implications for crisis management. Lack of empathy made policy makers blind to their adversaries's interests and to the pressures and constraints that shaped their leaders' decisions. Ignorance of war plans forced the leaders of Austria, Russia, and Germany to choose between inaction and an exaggerated military response. Their reluctant preference for the latter made war all but unavoidable.

My analysis reveals many disturbing parallels between 1914 and today, suggesting a disturbing conclusion: the chasm that so often separates military plans from political needs could once again become an important cause of war. The problem of empathy, an even greater threat to peace, I explore in a case study of the Cuban missile crisis. Lack of empathy, I argue, was a primary cause of the crisis and almost prevented its resolution. Both cases suggest the urgent need to make policy makers learn more about crisis management, their nation's strategic alerting procedures, and the interests of their adversaries.

Thresholds to War

Crisis escalation is a two-edged sword; it raises the risk of war in the hope of preventing it. By demonstrating willingness to wage war, leaders attempt to impress an adversary with their resolve and thereby encourage its leaders to moderate their behavior. But escalation often makes a crisis more difficult to resolve, because it increases for both sides the political costs of backing down.

When a state stipulates conditions under which it will go to war, it defines a threshold. Adversaries are unlikely to cross this threshold without carefully considering the consequences of their action. To do so represents a quantum leap in escalation; it signifies acceptance of war and all the uncertainty associated with it. Or, alternatively, it conveys disbelief in an adversary's resolve. Credible and clearly communicated thresholds have considerable deterrent value. The United States relies upon a carefully defined commitment of this kind to help secure its position in West Berlin. So does NATO; it is committed to go to war to repulse an attack against any of its members. More recently, the so-called Carter Doctrine attempted to define such a threshold with respect to Soviet-American competition in the Persian Gulf.[1]

Clear thresholds do not necessarily mean that lesser provocations will be tolerated, although they make it difficult to threaten war credibly with regard to them. Leaders often prefer ambiguity for this reason. Uncertainty can encourage restraint when leaders are concerned about unwittingly crossing their adversary's threshold. Ambiguous thresholds are most commonly used for this purpose where escalation has an asymmetrical impact upon the bargaining positions of the parties concerned. The side that is unable or unwilling to match its adversary's escalation may attempt to instill caution in its adversary by being deliberately vague about its own threshold to war.

States can also attempt to deter escalation by encouraging their adversary to believe that they will go to war in response to provocations that in fact they are prepared to tolerate. One example is

Khrushchev's threat in the 1962 Cuban missile crisis that "Soviet submarines would sink U.S. destroyers if they attempted to stop Soviet ships." On 26 October the *Marcula,* a Lebanese freighter under charter to the Soviet Union, was stopped, boarded, and inspected without provoking a Soviet military response.[2]

Thresholds are often ambiguous for another reason: because leaders do not know or do not wish to think about the circumstances in which they are prepared to go to war. For political or psychological reasons, they may shun precrisis discussion of this question. Ability to define thresholds can also depend on the political balance of power between "hawks and doves," the public's response to the prospect of war, or other conditions that become apparent only as the crisis unfolds. In the July crisis British leaders were able to commit themselves to the defense of France only after Germany had violated Belgian neutrality. German action galvanized British opinion in support of intervention, thereby shifting the balance of power within the cabinet.[3] Uncertainty also characterized the initial American response to the Soviet blockade of Berlin in 1948. At the onset of the crisis American officials were divided about how much opposition they should offer to a Soviet attempt to take over Berlin. Subsequently they began an airlift to provide essential foodstuffs and fuel to the beleaguered city, even though at first they were not convinced it would succeed. When the Soviets sent up barrage balloons to impede aircraft from landing, however, Truman made it clear to Moscow that the United States found the action intolerable. The balloons were withdrawn and no further attempt was made to interfere with the airlift.[4]

Thomas Schelling has analogized crisis to a variant of chess in which the possibility of mutual loss occurs whenever a queen and a knight of opposite colors cross the center line. The referee rolls a die. If an ace comes up, both players lose. If not, the game continues. "In this way," Schelling writes, "uncertainty imparts tactics of intimidation into the game. One can incur a moderate probability of disaster, sharing it with an adversary, as a deterrent or compellent device, where one could not take, or presuasively threaten to take, a deliberate step into certain disasters."[5]

Like most of Schelling's game theoretic analogies to international politics, this one is elegant, provocative, and misleading. Neither superpower has ever been so irresponsible as consciously to court nuclear war as a means of intimidating its adversary. This is true even of Cuba, the case from which Schelling derived his metaphor and so many of his other ideas about deterrence and compellence. Kennedy's strategy in that confrontation was guided by his commitment to find some means of demonstrating resolve while at the same time *minimizing* the risk of war. He settled on blockade as a result, even though he recognized that it would do nothing about the missiles already in Cuba. Advocates of the airstrike favored it not only because it would remove the missiles but because, they were convinced, the Russians would not respond militarily. In their view—rightly or wrongly—it was not a high-risk option.

Since Cuba the superpowers have been remarkably cautious. They have shied away from exercises in competitive risk taking because of their mutual fear of nuclear war. Such behavior conforms to one of the better-documented principles of cognitive psychology. Experiments show that most people are averse to risks; they are more sensitive to the possible costs of a gamble than they are to its payoffs. They shy away from gambles when they perceive possible costs as high.[6]

Schelling's analogy flies in the face of another established psychological principle. People have difficulty in tolerating ambiguity and uncertainty. When they take risks, they do their best to convince themselves that the risks involved are low or nonexistent. Statesmen are no different. Given this well-documented phenomenon, known as "bolstering," it is a mistake to speak of leaders consciously manipulating risks; more often they are trying unconsciously to dismiss them. Herein lies the principal reason for miscalculated escalation. Leaders who feel compelled by their domestic or foreign interests to pursue a foreign policy of confrontation are likely to convince themselves that the adversary will tolerate their initiative. Once committed to this course of action, they become to varying degrees insensitive to information or warnings that indicate their policy is likely to lead to war.[7]

Ambiguous thresholds encourage this kind of wishful thinking. They allow leaders to convince themselves more easily that their provocation will not elicit a violent response. The "forward policy" that India pursued in 1962, in its border dispute with China, is a case in point. Nehru and his foreign minister, Krishna Menon, supposed that China might offer some resistance to the patrols they sent into the contested territory, but they did not believe that these incursions would provoke war. Peking, for its part, tried to make clear that it would not tolerate the Indian policy, but it never specifically put India on notice that the patrols would lead to war. Nehru and Menon interpreted this absence as indicating lack of resolve. China's two-front offensive against the Indian Army took India's leaders entirely by surprise.[8]

Clear thresholds may be less likely to cause war by reason of miscalculation. But recent history indicates they are no guarantee against wishful thinking. A carefully delineated and communicated threshold can incorrectly be dismissed as a bluff, as it was in Korea in 1950. The People's Republic of China had put the United States on notice that it would intervene if non-Korean troops crossed the thirty-eighth parallel. American leaders refused to take this threat seriously; their army's advance to the Yalu River, Korea's border with China, led to war.[9] In 1967 President Nasser of Egypt may have misjudged Israel's well-publicized commitment to go to war in response to any blockade of the Straits of Tiran.[10] Complex political and psychological causes were in both instances responsible for misjudgment.

Thresholds to war can also be unambiguous *and* misleading. A hypothetical provocation might constitute a *casus belli,* but policy makers may simultaneously convey the impression to their adversary, by design or inadvertence, that it will not. The adversary may then escalate, convinced that its action will be tolerated by the other side, only to have it trigger war. This is the most dangerous kind of threshold.

Misleading thresholds can result from deliberate policy decision or from unintentional mishap. Policy makers may purposely mislead their adversary in the hope of provoking war—a charge unfairly leveled at the United States in the Cuban missile crisis and at Israel with regard to the origins of the 1967 Middle East war.[11] U.S. policy toward South Korea prior to 1950 is generally acknowledged to have been a case of inadvertence. American leaders may have unwittingly encouraged North Korea to believe that they would tolerate a northern invasion of the south.[12] Poor communication or inadequate policy coordination within a government can also be responsible for a misleading threshold. It happened in 1914, for example: German leaders misled their Russian counterparts about their country's likely response to Russian mobilization.

The July crisis is worth examining in detail because of the insights it provides into some of the causes of miscalculated escalation and loss of control. The two phenomena were so inextricably combined in that crisis that it is best to treat them together. As the case makes clear, political ignorance of military plans and their implications was an important catalyst of both sequences to war. Because of their ignorance, Russian leaders did not grasp the real meaning of mobilization; they also misjudged their ability to halt it once it had begun. The Germans, for whom Russian mobilization was a cause for war, failed to communicate this vital fact to St. Petersburg. Their failure can be traced to ignorance of German political leaders about the details of their country's war plan.

Miscalculated Escalation in 1914

Mobilization of any of the great Continental armies in the early 1900s set in motion a complex but carefully coordinated chain of events that culminated in the invasion of an adversary's territory. It was a time-consuming procedure, because, Germany aside, it required anywhere from ten days to two weeks between the order being given and an enemy's frontier being crossed in force. Without exception, the political leaders of the day labored under the misapprehension that statesmen could continue their search for a political solution as long as the mobilizing armies remained within their own borders. Luigi Albertini offers the judgment that

> Never, in tracing the history of the July crisis, can too much stress be laid on the point . . . that it was the political leaders' ignorance of what mobilization implied and the dangers it involved which led them light-heartedly to take the step of mobilizing and thus unleash a European war. The concatenation of Russian partial mobilization with Austrian general mobilization, of Austrian general mobilization with German mobilization, of French mobilization with German mobilization, was only perceived too late by all concerned. Nor did they realize that they must in any case adapt their actions to the existing plans of mobilization and that a plan of mobilization cannot be extemporized *ad hoc*. Likewise there was a general ignorance of the fact that mobilization rendered war inevitable beyond recall.[13]

Political leaders did not realize that military staffs everywhere were loath to stop mobilization once it had begun. The generals knew that a sudden halt would strand masses of men all along their country's railway lines, unserviced by any commissariat and unmovable until a new railway schedule was laboriously worked out. A mobilization so halted could leave a country more vulnerable to attack than it would have been in the absence of any mobilization at all. So it was highly unlikely that any of the Continental general staffs would have been willing to revoke their mobilization if they harbored even the slightest suspicion that their adversary was not prepared to do the same. The temptation to press on with one's own mobilization would have been very great indeed.[14]

Imposing political and psychological barriers also prevented people from revoking mobilization. Commitment has what Kurt Lewin called a "freezing effect."[15] Experimental evidence indicates that people, once they have made up their mind, are reluctant to reopen a decision in proportion to the difficulty they experienced in making it.[16] Before mobilizing, Austrian, Russian, and German leaders had gone through tortuous and emotionally wrenching deliberations. The commitment to mobilize, when finally made, brought a sense of relief, in some cases even euphoria, to the leaders involved. Subsequent appeals to them to reconsider their decision would have met with considerable opposition if not outright hostility. They would also have been resisted on political grounds. Any attempt to reconsider mobilization would have generated fresh and bitter struggles within Austrian, Russian, and German policy-making groups, something leaders were anxious to avoid. Any move toward peace would also have had to contend with the nearly universal outpouring of nationalist emotion triggered by mobilization and the prospect of war.[17]

Political leaders, ignorant beforehand of the almost irreversible organizational and psychological momentum associated with mobilization, were stunned by the reality. They also discovered, again belatedly, that they were quite wrong in their belief that they would have a significant grace period between proclaiming mobilization and the start of hostilities. Within hours of German mobilization, however, cavalry units crossed into Belgium to secure railheads essential to Germany for the invasion of France. Russian mobilization compelled German mobilization, and German mobilization constituted de facto war.

Sergei Sazonov, the Russian minister of foreign affairs, had no knowledge of the details of the Russian mobilization plan, let alone the German one. From the outset of the crisis he assumed that

Russia could carry out a partial mobilization, one directed only against Austria-Hungary. The Russian general staff did nothing at first to disabuse Sazonov of his illusions. The chief of staff, General N. N. Ianushkevich, was new at his job, ill-informed, and in any case not courageous enough to oppose Sazonov's desire for partial mobilization. The minister of war, General V. A. Sukhomlinov, it appears, was also ignorant of the details of Russia's mobilization plans. Only after Quartermaster General Yu. N. Danilov returned to Petersburg on 26 July did the general staff raise serious objections.[18]

Sazonov also believed that Germany, like Russia, would not actually go to war until it completed mobilization. He further assumed that both countries could refrain from war even after they had completed their respective mobilizations. When the czar ordered general mobilization on 31 July, Sazonov assured him that war could not start until the army received a special telegram, over the signature of the minister of war, indicating against which powers hostilities were to commence. All other measures were envisaged as merely preparatory.[19] The czar, acting in good faith, cabled the kaiser at 3 P.M. that afternoon: "It is technically impossible to stop our military preparations which were obligatory owing to Austria's mobilization. We are far from wishing war. As long as the negotiations with Austria on Serbia's account are taking place my troops shall not make any provocative action. I give you my solemn word for this."[20] On 1 August, learning of German military preparations, the czar once again telegraphed his cousin: "Wish to have the same guarantee from you as I gave you, that these measures *do not* mean war and that we shall continue negotiating for the benefit of our countries and universal peace dear to all our hearts."[21]

Russia's mobilization prompted a German ultimatum: Russia must revoke its mobilization. This the Russians found impossible to do. Germany mobilized, and Europe went to war. Some historians of the crisis, most notably Sidney Fay, have criticized the Russian mobilization as premature and for this reason hold Russian leaders largely responsible for the outbreak of war.[22] But this interpretation fails to take into account the fact that neither the czar nor Sazonov believed that their action would directly trigger war.

Sazonov's blunder was not entirely the result of Russian ignorance. Communications from foreign political leaders and diplomats had the effect of reinforcing Sazonov's belief that negotiations could continue after mobilization was completed. Sir Edward Grey, the British foreign secretary, was at first not at all upset by the Russian mobilization, because he viewed it as a deterrent to war.[23] The Germans, for their part, were rather vague about their likely reaction. On 26 July the German ambassador in London, Prince Lichnowsky, told Arthur Nicolson, permanent head of the Foreign Office, that "the Germans would not mind a partial mobilization . . . but could not view indifferently a mobilization on the German frontier."[24] Although a warning, this communication certainly did not put the Russians on notice that Berlin would view mobilization against Germany as a cause for war. Neither did Gottlieb von Jagow, the German secretary of state, when he discussed the problem with William Goschen, British minister in Vienna.[25]

The German chancellor, Theobald von Bethmann-Hollweg, seems to have been aware that German mobilization was tantamount to war, but he was not privy to the German war plan. He made only a belated and ineffectual effort to alert the Russians to the danger. He also failed to open Grey's eyes to this reality, even though he was well aware that Grey favored mobilization as a means of preserving the peace. On 26 July, Bethmann-Hollweg instructed Pourtalès, the German minister in St. Petersburg, to inform Sazonov that "preparatory military measures on the part of Russia directed in any way against ourselves would force us to take counter measures which would have to consist of mobilizing the army. Mobilization, however, means war, and would moreover have to be directed simultaneously against Russia and France, since France's engagements with Russia are well-known."[26]

Pourtalès did not grasp the import of the chancellor's message for, by his own account, in his subsequent conversation with Sazonov he failed to warn him that Russian mobilization would make war unavoidable. If anything, their talk seems to have strengthened Sazonov's impression to the contrary. Pourtalès relates that Sazonov put the question: "Surely mobilization is not the equivalent of war with you either. Is it?" The ambassador admitted that he *confirmed* Sazonov's belief to this effect, but he also claims to have warned him that "once the button is pressed and the machinery set in motion, there is no stopping it."[27] Even if we assume that Pourtalès made such a statement, which seems highly unlikely (his memoirs are riddled with self-serving falsehoods), he certainly did not know that German soldiers would enter Belgium within hours of the order to mobilize. If the German ambassador did not know, it is unreasonable to have expected Sazonov to have guessed the truth.

On 29 July, Bethmann-Hollweg dispatched a second communication on this subject to Pourtalès: "Kindly impress on M. Sazonov very seriously that further progress of Russian mobilization measures would compel us to mobilize and that then a European war could hardly be prevented."[28] Pourtalès reports he emphasized that the message "was not a threat but a friendly opinion," and Sazonov received it with "visible inward agitation." He replied only that he would communicate it to the czar.[29] Sazonov's agitation was caused by his belief that the Austrians were hell-bent on war with Serbia and trying to use Germany to keep Russia from interfering. To make matters worse, Russia's own generals were clamoring for mobilization, and neither Sazonov nor the czar had the courage to resist them. Bethmann-Hollweg's message, which once again failed to make the danger of Russian mobilization explicit, could not reverse the drift of Russian policy.[30] *Iacta alea est.*

The Causes of Civilian Ignorance

Our case study makes it apparent that European political leaders in 1914 shared two serious misconceptions about escalation: they believed that it consisted of discrete, reversible steps, and they believed that high levels of military preparedness in no way prevented diplomatic efforts to resolve crisis. We can attribute both misconceptions to the remarkably poor understanding by civilians of military affairs; political leaders were ignorant even about the mobilization plans of their own countries.

Why were the highest levels of government unfamiliar with the war plans? There were several mutually reinforcing causes. In the first instance ignorance was a result of the high turnover of officeholders. Most were not in office long enough to master the intricacies of military affairs or to establish effective control over the bureaucracies they nominally directed. There was also a widespread tendency in political circles to defer to military colleageus on what were considered technical questions, within the sphere of the general staff. Most political leaders had little or no inkling of the extent to which these "purely technical" matters would restrict their political freedom in a future crisis.

Armies and navies, of course, did nothing to encourage greater civilian understanding of military affairs; quite the reverse. Military leaders everywhere did their best to keep their civilian colleagues in the dark about the details of their respective war plans and associated requirements. They volunteered information only to the extent that it was necessary to secure political support for their budgetary, manpower, or related administrative needs. When compelled to testify before cabinets or parliaments, they generally provided political authorities with information, often information they knew was inaccurate, that justified their strategy or the manpower and material requirements they set for it.[31]

Between 1880 and 1914 several European countries made unsuccessful attempts to reorganize their armed forces, primarily to bring them under greater civilian control. Political leaders in Britain, France, and Russia also attempted to force new weapons upon reluctant services that viewed innovations as threats to cherished traditional values. Continental cavalries and the British Navy were notorious for their resistance to progress.[32] Military secrecy therefore became an important means by which generals and admirals sought to protect their institutions. In many countries that secrecy was also symptomatic of deeper political divisions.

In the course of the nineteenth century economic and scientific progress, and the cultural and political ideas associated with it, had transformed Europe. Change led in varying degrees to shifts in political power, away from the landed gentry and toward the emerging commercial and industrial classes. Political participation also became more widespread. In western Europe, where political change had gone furthest, armies remained the public institution least affected by democratization. They were self-consciously the last redoubt of aristocratic classes fighting rearguard actions to preserve their dwindling prerogatives.[33]

In Germany the conflict between the army and the new classes was particularly acute.[34] Moltke and other military leaders described it as a clash between *Händler und Helden* (hawkers and heroes) for the very soul of the nation.[35] Their distrust of political authorities, even those from their own class, was pronounced. So was their unwillingness to take civilians into their confidence, something that in any case they had little need to do. Bismarck had successfully defeated the attempt by the National Liberals to impose parliamentary control over the army during the prolonged constitutional crisis of the 1860s. The constitution he subsequently drew up for the Reich had left the military, their budget aside, entirely independent of civilian control. Officers swore fealty to the kaiser, and the military chiefs reported directly to him, not to his civilian ministers.[36]

The pyramid of German political-military policy making also left the services relatively uninformed about each other's war plans. They jealously, and on the whole, successfully, guarded their prerogatives and operational independence from each other. The navy, for example, repeatedly rejected the general staff's request to share intelligence.[37]

Such secrecy served important organizational ends. In a highly centralized and compartmentalized policy-making environment, information was a source of power. A service's monopoly over information relevant to its sphere of activity permitted its chiefs to manipulate the kaiser's and chancellor's understanding of military problems in accord with institutional preferences. Kaiser, chancellor, and foreign ministry alike were for this reason vouchsafed only the most superficial description of the Schlieffen Plan. They knew that it involved an invasion of France through Belgium and a holding operation in the east against Russia. Of the details of the plan, however, they knew next to nothing. Bethmann-Hollweg could in all honesty mislead the Russians into believing that their mobilization, while highly threatening to Germany, did not of necessity mean war.

Contemporary Civil-Military Relations

Is there any contemporary resemblance to 1914? Klaus Knorr, a scholar with considerable governmental experience, is emphatic that the two periods are not analogous. "Today's political authorities and publics," he writes, "are more knowledgeable about the implications of existing military technologies, and are less disposed to let the military have their way, than was the case before 1914."[38] Knorr is not alone in asserting that widespread recognition of the dangers of developing strategy in a political vacuum has led to efforts in the United States, and probably in the Soviet Union as well, to impose political criteria on military planning.

The other big difference between 1914 and today is the absence of the class divisions within countries which were a principal cause of civilian-military conflict in Germany, France, and Britain in the early years of the century. The Soviet and American military establishments are certainly distinctive institutions, with their own traditions, outlooks, and sense of special mission. But neither constitutes a state within a state, nor an institutional base for political dissidents scheming against the established regime. At most we can say that both military organizations are repositories of more traditional and nationalist values, orientations that do not necessarily clash with the notion of civilian control. In comparison to many other great powers, past and present, both superpowers have succeeded, by quite different means, in establishing and maintaining clear political control over their respective militaries.[39]

These differences are important, but there are also some striking parallels to 1914. The most significant is an ingrained institutional reluctance on the part of both the American and the Soviet armed forces to share important military information with their political overlords. Just like their 1914 predecessors, they have sought, generally with success, to use their monopoly over the details of strategy to ensure their operational control over the means of war.

In the United States efforts by the Strategic Air Command (SAC) to safeguard its operational independence are well documented. In the late 1940s and early 1950s SAC fought successfully to gain control over nuclear weapons and the strategies for employing them. Landmarks in this struggle were Eisenhower's 1953 decision to make atomic bombs immediately available to the military and the creation in August 1960 of the Joint Strategic Target Planning Staff (JSTPS) at Offut Air Force Base in Nebraska. JSTPS represented something of a bureaucratic compromise. Two years earlier the SAC Commander, General Thomas Power, had proposed that he be given sole authority over all strategic targeting. But Secretary of Defense Thomas Gates insisted that coordinated, central war planning was essential to the national interest, and he established JSTPS with an admiral as its vice-director. In practice, however, the location of JSTPS at Offut, an air force facility well away from prying eyes in Washington, gave SAC a relatively free hand in the preparation of targeting lists and options.[40]

Throughout the 1950s SAC prepared to fight its own kind of war: a preemptive, all-out attack against the Soviet Union, Eastern Europe, and the People's Republic of China. SAC's approach reflected General Curtis LeMay's view of the purpose of military force. "I'll tell you what war is about," he once lectured nuclear weapon designer Sam Cohen; "you've got to kill people and when you've killed them enough they stop fighting."[41] As open as LeMay and his successors were about the philosophy of SAC's war plan, they were secretive about its details. In 1960, for example, a three-man team that Eisenhower dispatched to Offut, headed by presidential science adviser George Kistiakowsky, was given the royal run around.[42]

The Kennedy administration at first fared no better. The JCS discouraged the president from looking into the SIOP; it was all "too technical," they said, "routes, refueling schedules, target coordinates, overlapping E-95 circles and the like." SAC for its part protested (on grounds of security) efforts by McNamara to send Defense Department officials to Offut to study the war plan, although McNamara himself was invited for a visit. Under pressure, they finally agreed to reveal details to a small team of McNamara aides which included Daniel Ellsberg. McNamara subsequently insisted on the development of a more flexible war plan.[43]

The second SIOP, approved by McNamara in June 1962, represented the first attempt to make nuclear strategy from the "top down" rather than the "bottom up."[44] All of McNamara's successors have pursued this goal; they have instituted mechanisms that have succeeded in making SAC and the nuclear navy more responsive to civilian directives. It would nevertheless be wrong to conclude that the civilian-military misunderstandings that contributed to the outbreak of war in 1914 could not occur again. The strategic bureaucracy remains difficult to govern for many structural reasons. As Paul Bracken aptly observes,

> It lacks flexibility, has rigid goals, and adheres to "legitimized" facts which are often of questionable or arbitrary origin. Most important, a strategic machine . . . is opaque to outside political control; there is a disconnect between cause and effect for purposes of political com-

> mand and control over it. It is not even transparent to those who work inside of it because of the highly specialized nature of the work (and also because of the rigid security compartmentalizations in effect).[45]

Beyond these organizational impediments to political control, the military has strong institutional incentives to resist detailed civilian knowledge of how alerts work and how targets are selected. Since the experience of Vietnam, all of the services have been fearful of civilian "micromanagement" of military operations, something most men in uniform believe can only mess them up.[46] I know from my own experience as scholar-in-residence in the Central Intelligence Agency that the military penchant for secrecy can sometimes go to absurd extremes. CIA efforts to model the outcome of strategic exchanges between the superpowers were complicated by the unwillingness of the air force and the navy to provide relevant information about U.S. weapons. My colleagues and I sometimes felt we knew more about the Soviets' force capabilities.[47]

General William E. Odom, who helped shape the Carter administration's strategic policies, confirms that this problem continues to exist at the highest levels of government. According to Odom, one of the motives behind Carter's promulgation of PD-59 was the desire to gain more control over targeting policy. Other Carter officials have testified to the reluctance of the JCS and SAC to provide any of the supporting detail of the SIOP to the White House. Said one Defense official: "They're not going to tell anybody where the airplanes are going to refuel . . . [or] why it's a good idea to hit tank parks even after the tanks are gone."[48]

Civilian Escape from Responsibility

Civilian frustration with the military has been matched only by military frustration with the civilians. Military officers, although secretive about the details of the SIOP, have since LeMay's time actively sought political guidance about what kind of war they should plan to fight. They have not always been successful in getting the guidance they want. In 1961 the SIOP was rewritten in response to guidelines established by then secretary of defense

Robert McNamara. But for the next decade war planners at Offut received no further instructions, although more nuclear weapons entered the American arsenal during these years than at any time before or since. General Richard H. Ellis, former director of strategic target planning, reports that the officers involved in updating the SIOP used to read speeches by the president and secretary of defense in the hope of finding some hints about what to do with all the new weapons.[49]

In January 1974 Richard Nixon signed National Security Decision Memorandum 242, known ever since as the Schlesinger Doctrine. It provided new guidance for target selection, stressing the need for options that would include the limited use of nuclear weapons against specific military targets. Jimmy Carter's Presidential Directive 59 and Ronald Reagan's National Security Decision Directive 13 have continued the practice of providing high-level guidance to war planners. The Carter and Reagan directives, each only a few pages long, were used by the Office of the Secretary of Defense to prepare a "Nuclear Weapons Employment Policy," which went to the Joint Chiefs of Staff, who in turn prepared a more elaborate document based on it for use by JSTPS.[50]

Despite these improvements in target guidance, civilian understanding of the SIOP, its options and implementation, and their political implications, continues to be meager. The SIOP itself is stored in a computer; hard copy would consist of a five-foot-high pillar of printout. Even the briefing book prepared and periodically revised for high-ranking political officials is several inches thick. Of necessity, presidents must rely on well-informed advisers to pass judgment on something so complex and technical. They must be content with a broad overview of the SIOP and their responsibilities in its execution. But with the exception of Jimmy Carter, no president has given any evidence of having this much mastery of his country's war plans and his own role in them. Richard de Lauer, former under secretary of defense for research and engineering, reports that one of the things that has most troubled defense officials over the years "is the difficulty of getting the President to sit down to practice."[51]

Eisenhower was greatly interested in the military logic being developed by the air force to rationalize the new targeting plan but had little interest in familiarizing himself with the details of the plan or his role in it. Kennedy devoted no more time to the second

SIOP. Lyndon Johnson and Richard Nixon were uninterested in war plans, and they were reportedly quite impatient during their briefings.[52] In his memoirs retired admiral Elmo Zumwalt describes a National Security Council meeting in late January 1974, making it apparent that Nixon had absolutely no idea of what NSDM 242 meant even after he had signed it.[53] Gerald Ford was equally ill-informed, and Ronald Reagan has carried on the tradition.[54] Only Jimmy Carter showed a real interest in nuclear questions and made some effort to familiarize himself with the details of nuclear strategy. He was also the first president since Eisenhower to authorize a command post exercise. Unlike Eisenhower, he took part in it to test and familiarize himself with evacuation and authorization procedures.[55]

What accounts for this striking lack of presidential interest in nuclear strategy? The most likely explanation is psychological. People shun anxiety-provoking subjects and decisions, and no responsibility is more likely to arouse anxiety in a president than contemplation of his role as commander-in-chief of the armed forces in the event of nuclear war. As even the least well-informed president knows, it would be up to him to decide if and when to order a nuclear attack, an action that could wipe out more people than Adolf Hitler and Genghis Khan combined. Almost all of these fatalities, moreover, would be perfectly innocent folk who had the misfortune to live in the wrong political jurisdiction. The fact that such an attack would presumably be ordered only in retaliation would probably do little to diminish the stress and moral burden any president would feel in contemplating sending the go-code. The alternative, no retaliation, would be equally unpalatable for quite different reasons.

The president in a nuclear conflict would appear to be in an extreme example of a "no-win" situation. The personal, political, and human costs of any of the choices open to him would be truly monumental. This kind of dilemma encourages defensive avoidance; the person concerned tries, if at all possible, not to think about the problem.[56] And so presidents, with one exception, have sought to avoid questions of nuclear strategy, the SIOP, and their role in implementing it. For the same reason, they may also have minimized the probability of a nuclear war between the superpowers.[57]

Defensive avoidance could account for John Kennedy's incredulity when one of his advisers suggested during the Cuban missile crisis that he sleep in the White House bomb shelter.[58] It could also explain why Richard Nixon's mind was reported to wander when the SIOP was discussed, and why Ronald Reagan is alleged to have postponed his SIOP briefing for three years after assuming office. It may even help to explain why Reagan has been attracted to the Strategic Defense Initiative, something that seems best understood as an escape from nuclear reality. The fact that Reagan's unexpected commitment to it came hard on the heels of his SIOP briefing, an encounter that reportedly left him ashen-faced and speechless, lends credence to this speculation.

What is true for presidents also holds true for their advisers. For the most part, they too have avoided confronting the details of nuclear operations.[59] Joseph Nye, Jr., who has conducted interviews with top politically appointed government officials, was shocked by their ignorance. "Most of these officials," he reported, "know almost nothing about what they would do in a nuclear crisis." Nor do they have time to learn. Trying to teach them, Nye lamented, has been "like force-feeding them with a hose."[60] The situation at the Defense Department is no better; there Caspar Weinberger is reported to have left the matter entirely to his military aides.[61]

The fact that the White House and the office of the secretary of defense have pushed nuclear strategy and crisis management into the background raises few objections from the military who, we have seen, have their own reasons for discouraging civilian interest in nuclear planning and operations. Reinforcing incentives thus contribute to mutual ignorance: civilians have little understanding of military operations, while the officers who plan these operations know little about the options and related procedures political leaders would find helpful.

The limited evidence we possess suggests there is good reason to believe that the same problem exists in the Soviet Union; if anything, it is likely to be more serious there. Compartmentalization and secrecy are the very hallmarks of the Soviet political system and are most pronounced in the realm of national security. Timothy J. Colton, an authority on Soviet civil-military relations, reports the Soviet officer corps functions behind a high wall of secrecy. It possesses, together with its allies in the defense industry, a "near monopoly" on information germane to making military decisions.[62]

Arms control negotiations and more informal contacts over the years have also made it clear that the Soviet military shares less information with political officials than does its American counterpart. During the SALT I talks General Nikolai Ogarkov, the chief of staff's representative on the Soviet delegation, took aside his American opposite number, General Royal Allison, and berated him for discussing details of Soviet military operations that "need not concern his civilian colleagues."[63] Such secrecy apparently exists within the military itself. As high-ranking an official as Major General M. Gorianov of the Engineering Technical Service had to use published Western sources on Soviet nuclear weapons capabilities for a top-secret 1960 article, because his colleagues denied him access to official data.[64]

Information about nuclear weapons and strategy appears to be controlled very tightly by the Main Operations Directorate of the Soviet General Staff. This directorate functions without any immediate civilian supervision, because the Central Committee lacks an autonomous department to oversee this policy area.[65] Also absent are civilian strategic advisers, think tanks, and academic experts. None of these civilian influences on strategy, so important in the United States, exists in the Soviet Union.

Some evidence also indicates that Soviet leaders may be just as unwilling as their American counterparts to involve themselves in the details of nuclear strategy or to take seriously the possibility of nuclear war. Nikita Khrushchev confided to Egyptian journalist Mohamed Heikal: "When I was appointed First Secretary of the Central Committee and learned all the facts about nuclear power I couldn't sleep for several days." He went on to confess that he finally convinced himself "that we could never possibly use these weapons, and when I realized that I was able to sleep again."[66]

The Implications of Civilian Ignorance

Political ignorance in the United States and the Soviet Union of the details of war planning does not bode well for crisis management. It threatens to make political leaders the captives of prepackaged military options, just as it did in 1914. To make matters worse, time constraints now would most likely be measured in hours instead of in days; there would be even less of a chance for improvisation today than there was in the July crisis.

In July 1914 the military's refusal to deviate from existing war plans had an important psychological component. Resistance to change correlates positively with stress. When stressed, people tend to resist innovation and seek security in familiar routines.[67] Moltke was under great stress throughout the crisis. He was, according to Corelli Barnett, an "old and desperately tired man, his physical and mental powers spent by a month of crises. . . . His plump features pallid with fatigue and unstrung nerves, Moltke could barely drive his own body through the daily routine of duty, let alone drive his armies through catastrophic danger into victory."[68] Moltke sought strength in the Schlieffen Plan; it assumed something close to totemic significance for him. Victory, he assured the kaiser and chancellor, was possible only if the plan was faithfully executed. Any deviation, he warned, was certain to result in disaster.[69]

Nor did the statesmen of 1914 simply find themselves captives of military plans. Only belatedly did they discover just how unsuitable these plans were to their political needs. Political leaders in Austria, Russia, and Germany were forced to choose between an inappropriate military response and no military response at all.

Russian and German leaders in 1914 were caught in a trap their military planners had unwittingly laid for them years earlier. Russia mobilized in response to Austria's declaration of war on Serbia, an act St. Petersburg would not let pass unchallenged. The czar preferred a "partial mobilization," that is, mobilization in only the four military districts facing Austria-Hungary. The general staff successfully opposed the idea; staff officers insisted that partial mobilization was impossible because there were no plans for it.[70] Russian mobilization triggered German mobilization, because the Schlieffen Plan was predicated on Germany's need to defeat France before the slowly mobilizing Russians advanced too far in the east. The Schlieffen Plan also committed Germany to attack Belgium, which analysts deemed necessary in order to secure enough maneuvering room for the invasion of France. This decision made it all but inevitable that Britain would enter the war on the side of the Entente. The mobilization plans of the powers, as Luigi Albertini so aptly put it, represented "masterpieces of military science but also monuments to that utter lack of political horse sense which is the main cause of European disorders and upheaval."[71]

Do contemporary war plans suffer from the same failing? Technically, they too are "masterpieces of military science." They are extraordinarily complex blueprints for action, which attempt to coordinate the panoply of strategic forces at the command of the superpowers in the most efficient manner. However, on the few occasions that we know political leaders actually looked at these plans, the politicians found them entirely unsuitable to the ends they sought.

The Kennedy administration's dissatisfaction with military planning prompted a total overhaul of the first SIOP. In July 1961 McNamara's aides discovered that the existing war plan called for the United States to launch an all-out nuclear attack against the Soviet Union in the event that Russian troops turned back an allied probe toward Berlin. The special assistant for national security affairs, McGeorge Bundy, advised the president "that the current strategic war plan is dangerously rigid" and "may leave you with very little choice as to how you face the moment of nuclear truth." The war plan, according to Bundy, was deliberately designed "to make any more flexible course very difficult."[72]

Subsequent SIOP revisions, most of them designed to incorporate more flexible options, still confront policy makers with little more than a choice among a small number of preplanned options. A major reason is the complexity of any attack plan and the corresponding need to coordinate in detail all of the operations it involves. Without computer-generated timing schedules, warheads delivered by different kinds of weapons systems will not reach their targets in proper sequence. They may interfere with one another, possibly causing U.S. missiles to destroy U.S. bombers instead of destroying their assigned targets. These schedules cannot be reworked on the spur of the moment. On this point General Thomas Power, former commander of SAC, is as adamant as Field Marshal Moltke was before him. "You cannot coordinate a plan," he insists, "after you have been told to go to war. It all has to be part of a well-thought-out, well-worked-out plan. And there is one basic law you must follow. Do not change it at the last minute."[73] The import of his remarks is clear: like it or not, American leaders are stuck with the existing war plans; they cannot tailor them to the political needs of the moment.

How useful are these prepackaged options? Greater political input into the planning process has not, it appears, resulted in war plans that political leaders would find any more helpful in a crisis. Henry Kissinger reports that his own doubts about the wisdom of war-fighting strategies, even limited ones, were confirmed in the aftermath of the Arab-Israeli war. He discovered that the most limited military action envisaged by Pentagon planners in response to a Soviet invasion of Iran entailed pummeling the Soviet Union and its forces entering Iran with more than two hundred nuclear weapons. Kissinger's objection that this would constitute full-scale war, not just a signal of American resolve, elicited an alternative suggestion from the Joint Chiefs: they would use only a couple of nuclear weapons. Kissinger found this option equally unpalatable for just the opposite reason.[74] The current administration would probably be just as appalled as its predecessors by the inappropriateness of the nuclear options available to it—if it made the effort to examine them.[75]

The situation is probably no different in the Soviet Union. Soviet specialists are in general agreement that Soviet consideration of nuclear war would take place in a setting heavily influenced by a military committed to a preemptive doctrine. Stephen Meyer warns that "the lack of alternative military-strategic advice implies that Soviet political leaders would be unprepared for policy innovation under stress and susceptible to professional military influence. These signs all point toward greater dependence on prior planning, with less ability or tendency to reach out for creative alternatives."[76]

Soviet war plans could also confront Soviet political leaders with the same awkward choice between doing too much and doing too little. Soviet conventional forces in Europe are geared for an all-out offensive.[77] As Jack Snyder observes,

> On the surface, it may seem that the awe-inspiring Soviet military machine and its intimidating offensive doctrine are apt instruments for supporting a policy of diplomatic extortion. It may, however, pose the same problem for Soviet statesmen that the Schlieffen Plan did for Bülow and Bethmann. Soviet leaders may be self-deterred by the all-or-nothing character of their military options. Alternatively, if the Soviets try to press ahead with a diplomacy based on the "Bolshevik operational code" principles of controlled pressure, limited probes

and controlled, calculated risks, they may find themselves trapped by military options that create risks that cannot be controlled.[78]

The Soviet Union appears committed to a policy of avoiding nuclear war but of using nuclear weapons on a massive scale if that policy fails. Soviet military doctrine denies the political and military feasibility of limited nuclear war; it stipulates that nuclear weapons, if used, should aim to disrupt or prevent coordinated military activity by the enemy. According to Benjamin Lambeth, "the current U.S. development of selective nuclear options has no analogue in known Soviet military concepts and contingency planning, which appear predominantly oriented toward massive nuclear strikes aimed at capitalizing on the inherent power of the initiative and achieving some recognizable form of military victory."[79]

Some Western commentators have speculated that Soviet targeting strategy may not be a faithful reflection of Soviet doctrine. They think that the Strategic Rocket Forces may have developed a range of limited nuclear options but that Soviet spokesmen maintain their public commitment to an all-out war strategy for purposes of deterrence.[80] A discrepancy of this kind between declared and actual policy is entirely possible. The United States pursued such a two-track policy during the McNamara years, when it was publicly committed to the countervalue targeting strategy associated with mutual assured destruction but secretly worked to develop more limited counterforce options and the weapons necessary to implement them.[81] There is no hard evidence that the Soviets have developed limited options, however, and much reason from past experience to suppose that Soviet doctrine is an accurate guide to military planning. If so, in some future crisis Soviet leaders, even more than their American counterparts, could find their existing military options entirely unsuitable to their political needs. This discovery could prompt caution, but, as in 1914, it could also push Russian leaders, in circumstances where they could not accept inaction, toward a more extreme response.

Perhaps the greatest danger associated with political ignorance of the details of war planning is conceptual. There is a widely shared notion in the United States that crisis management consists of discrete, reversible steps that one takes up the rungs of some metaphorical ladder of escalation in order to demonstrate resolve and thereby moderate an adversary's behavior. Belief in the inherent *manageability* of escalation derives in large part from the experience of the Cuban missile crisis as interpreted by the spate of journalistic, participant, and scholarly analyses that followed it. But as Chapter 3 demonstrated, the "lessons of Cuba" would be entirely inappropriate to a future superpower crisis. They are not only inappropriate but, to the degree they encourage in political leaders a false confidence about their ability to control escalation, dangerous.

The popular image of the Cuban missile crisis is also misleading. It stresses Kennedy's wisdom and resolve and downplays his uncertainties and fears. In point of fact Kennedy found the crisis a sobering experience; he came to realize just how many things could go wrong, and he considered that he and Khrushchev were lucky to avoid war. Kennedy is the last president to have managed an acute superpower crisis. His experience, and some of the more recent academic literature that emphasizes the difficulties of controlling large political and strategic bureaucracies, remind us of the ways in which the reality of crisis management differs from public conceptions of it. Unfortunately, this lesson of Cuba seems to have made little impact.

One reason is the extensive turnover of officials at the highest levels of government every time a new administration comes to power. To make matters worse, incoming officials rarely meet or speak at any length with their outgoing counterparts. Often they also find the file cabinets empty, because their predecessors have all too frequently departed with their papers in the expectation of writing memoirs.[82] Such circumstances preclude the development of any kind of institutional memory. The academic literature, which might to some extent bridge the gap, has apparently failed to make any impact on official thinking. Political leaders and their advisers have neither the time nor the inclination to read it. The only other antidote to unrealistic thinking about crisis management is familiarity with the details of military plans and operations. First-hand experience would drive home the lesson of just how difficult it is to control escalation. But important practical and psychological barriers prevent political leaders from attaining this knowledge.

Motivated Bias and Miscalculated Escalation

Like all decision makers, political leaders are emotional beings, not rational calculators. They are beset by doubts and uncertainties, they struggle with incongruous longings, antipathies, and loyalties, and they are reluctant to make irrevocable choices. Important decisions generate internal conflict, which takes the form of simultaneous tendencies to accept and to reject a given course of action.

The stress associated with making a decision can become acute when a policy maker concludes that a real prospect of serious loss is associated with any open course of action. In these circumstances the policy maker will become emotionally aroused and preoccupied with finding a less risky policy alternative. If, after further investigation, the official concludes that it is unrealistic to hope for a better strategy, he or she will terminate the search despite continuing dissatisfaction with the available options. The result is a pattern of defensive avoidance.[83]

Defensive avoidance can come in three forms, according to Irving Janis and Leon Mann: procrastination, shifting responsibility, and bolstering. The first two are self-explanatory. Bolstering is an umbrella term that describes various psychological tactics designed to allow policy makers to entertain expectations of a successful outcome. The policy maker turns to bolstering tactics after losing hope of finding an altogether satisfactory policy option and when unable to postpone a decision or foist the responsibility for it onto someone else. Instead, the official commits himself to the least objectionable alternative and proceeds to exaggerate its positive consequences or minimize its negative ones. Some officials may also deny the existence of aversive feelings, emphasize the remoteness of the consequences, or attempt to minimize responsibility for the decision after it is made. The policy maker continues to think about the problem but wards off anxiety by practicing selective attention and other forms of distorted information processing.[84]

Bolstering can serve a useful purpose. It helps a policy maker forced to settle for a less than satisfactory course of action to overcome residual conflict and move more confidently toward commitment. But bolstering has detrimental consequences when it occurs before policy makers have made a careful search among the alternatives. Then it lulls officials into believing they have made a good decision when in fact they may have avoided making a vigilant appraisal of the possible alternatives, in order to escape from the conflict such an appraisal would engender.[85]

German policy making in 1914 suffered from this problem. Germany's political leaders believed it was essential to support Vienna's effort to subjugate Serbia in order to preserve their alliance with the Dual Monarchy. They also wanted to avoid provoking a major European war.[86] In point of fact these were contradictory objectives. Nevertheless German leaders convinced themselves that the other powers would stand aside while Austria crushed Serbia. This motivated bias was also responsible for the failure of German leaders to realize the extent of their miscalculations as the crisis unfolded. Evidence to this effect was either ignored, misinterpreted, or suppressed until the very denouement of the confrontation.

Wishful thinking was not unique to Germany. Austrian leaders also resorted to perceptual sleights of hand to make reality consonant with their political-military needs. Their dilemma derived from the prospect of a two-front war, against Serbia in the south and Russia in the northeast. The Austrian general staff had calculated it needed twenty divisions to conquer Serbia. They would be left with only twenty-eight divisions to defend against the fifty-three that Russia could throw into Galicia within a month of mobilization.

If the Austrians sent more troops to the Russian front, however, they would no longer have sufficient forces to invade Serbia. Unwilling to contemplate this prospect—Austrian leaders believed that Serbia's subjugation was the only means of coping with their empire's nationality problem—they convinced themselves that German backing would deter Russia from going to war. They also took refuge in the belief that Russian entry into the war would *still* not interfere with their conquest of Serbia, because the Russian divisions that could be used to attack Austria would be needed further north to defend against Germany.[87]

So essential was a German invasion of Russian-occupied Poland to Austria's plans that Austrian political and military authorities convinced themselves the Germans would undertake a major offensive in the east. In 1909, in response to Austrian pleas, Moltke committed himself to some kind of offensive against Russia. The Austrians clung to his promise despite all the subsequent reports they received indicating that Germany would make its offensive effort in the west instead, against France. German staff officers, aware of Austria's misconception and of the reasons for it, even advised their Austrian colleagues not to count on a major German offensive in the east. Their warnings, and there were several, aroused such a storm of protests from Vienna that German generals made no further attempt to disabuse the Austrians. To have done so, after all, would have weakened Austrian resolve to act decisively with Serbia, and it might also have led Austrian authorities to question the value of their alliance with Germany.[88]

In Russia competing and largely contradictory military requirements were also "resolved" by wishful thinking and denial. To placate pan-Slav sentiment, the general staff was compelled to plan for an offensive against Austria in Galicia. But strategic considerations dictated an offensive in the north, into Prussia, in order to relieve German pressure on France. The general staff knew that two offensives were well beyond the Russian Army's capability; some staff officers even doubted whether it could carry out a successful offensive on one front. The generals were nevertheless divided in their opinion about which offensive to forgo. Unable to strike a compromise, they committed Russia to less grandiose efforts on both fronts, making it highly unlikely that either offensive would succeed.[89]

Recognition of their unenviable military situation made Russia's leaders understandably anxious to avoid war. At the same time, however, domestic political pressures and concern for their standing as a great power made it impossible for them to tolerate Austrian subjugation of Serbia.[90] The only way out of their dilemma would have been some diplomatic solution to the crisis which protected Serbian independence. Perhaps this position may help to explain why Sazonov was so willing to believe that Russian mobilization would act as a deterrent and encourage Germany to rein in Austria.

In all three countries, political needs encouraged unrealistic expectations regarding the behavior of friends and foes alike. These illusions permitted leaders to finesse the hard choices that their respective national interests required but that they were unprepared—politically, organizationally, and psychologically—to make. They also allowed them to preserve their peacetime equanimity. In crisis, however, such illusions were dysfunctional, because they made the leaders who found refuge in them grossly insensitive to their adversaries' interests and their own military limitations. This double blindness led them to underestimate adversarial resolve and overestimate their own bargaining advantages. It was a prime cause of the miscalculated escalations that led to war.

World War I may be the most dramatic example of how wishful thinking contributed to war, but it is certainly not the only one. More recent manifestations of the same phenomenon include the American decision to cross the 38th parallel in Korea, the Sino-Indian border war of 1962, the Middle Eastern wars of 1967 and 1969, and the 1982 Falklands War. In all of these cases, clashing and seemingly irreconcilable threats were responsible for miscalculations that resulted in war.[91]

Cuba: A Case Study of Miscalculated Escalation

No event of the Cold War brought the superpowers as close to blows as did the Cuban missile crisis. It was a striking example of miscalculated escalation; for whatever reason, Khrushchev wrongly concluded the United States would tolerate the furtive emplacement of nuclear-armed missiles in Cuba. Fortunately Soviet leaders realized the extent of their miscalculation in time to back off and preserve the peace.

Soviet leaders were not alone in misjudging their adversary's willingness to tolerate provocation. Many of Kennedy's advisers might have done the same thing; they were certain that the Soviet Union would tolerate an American airstrike against the missile bases in Cuba. The president, however, sensitive to Khrushchev's need to save face, opted for a blockade. The crisis, although resolved, gives vivid testimony to the difficulty the superpowers have in understanding and predicting each other's behavior. This difficulty continues; and so the Cuban experience remains a sobering lesson for the future.

The View from Moscow

Early in the autumn of 1962 President Kennedy responded to reports that the Soviet Union was secretly deploying missiles in Cuba by issuing a series of stern warnings to Moscow.[92] Kennedy's warnings were also communicated to Moscow through private channels, in order to reinforce them and to dissuade the Soviets from dismissing them as mere campaign rhetoric.[93]

Soviet statements, public and private, appeared to indicate Moscow recognized the gravity of the American warnings. Graham Allison ventures the judgment that they can be interpreted only as assurances to Kennedy that Moscow understood and respected American strategic and political interests in Cuba:

> The United States formulated a policy stating precisely "what strategic transformations we [were] prepared to resist." The Soviet Union acknowledged these vital interests and announced a strategy that entailed no basic conflict. This would also seem to be a model case of communication, or signaling, between the superpowers. By private messages and public statements, the United States committed itself to action should the Soviets cross an unambiguous line (by placing offensive missiles in Cuba). All responses indicated that the Soviets understood the signal and accepted the message.[94]

Moscow, it now seems evident, had decided sometime in June 1962 to put missiles in Cuba. Russian assurances to the contrary were designed to lull the Kennedy administration into inaction until the missiles were actually deployed. How could Soviet leaders have embarked upon a course of action almost certain to provoke a major confrontation with the United States whether or not the missiles were discovered before they became operational? More troubling still, why did Moscow continue to proceed with its strategic ruse even after Kennedy's several warnings, stepped-up American surveillance, and well-advertised military preparations had clearly indicated that confrontation was a near certainty?

By all accounts, Soviet leaders had strong, even compelling, incentives to put missiles into Cuba. Today the most widely accepted interpretation attributes their action to the Soviet need to redress the strategic balance. In the autumn of 1961 the Kennedy administration decided to tell Moscow it knew that the Soviets' first-generation ICBM had proved a failure. It did so in the hope of moderating Khrushchev's bellicosity over Berlin. But the revelation also put Moscow on notice that the United States, through its satellite reconnaissance, realized the full extent of Soviet vulnerability to a first strike. Missiles in Cuba, many analysts therefore argue, were conceived of as a "quick fix" to compensate for Soviet strategic inferiority.[95]

Graham Allison believes the missiles could have appealed to influential factions or institutions within the Soviet hierarchy as a possible solution to problems they confronted. The missiles offered strategic planners a readily available means of coping with American strategic superiority. The foreign ministry might have been attracted to the venture as a means of dramatizing Soviet support for Castro. The deployment would also demonstrate Soviet resolve to Peking, then contesting Moscow's leadership of the communist movement. If the ploy succeeded, it would give Khrushchev more "chips" to use in negotiations over Berlin. Finally, an aggressive policy promised to advance Khrushchev's political interests and those of his supporters, who must certainly have felt the need for a major success after the failure of their two Berlin offensives and their domestic agricultural programs. For all of these reasons, Allison speculates, a powerful coalition may have developed in favor of putting missiles into Cuba—a coalition composed of political leaders and bureaucrats who saw the move as beneficial to their parochial interests.[96]

To the extent that Khrushchev and those around him felt a need to deploy Soviet missiles in Cuba, they also felt a need to believe they could do so successfully. Many of the explanations that have been offered to account for Moscow's misjudgment are perhaps better understood as rationalizations, invoked by Soviet leaders to justify their decision and to dispel whatever anxiety they may have harbored about the risks it entailed. After choosing the initiative, they became insensitive to information that suggested it would not succeed.[97]

The timing of the Soviet decision was probably very important in this regard. All the available evidence points to the decision's having been made sometime in the late spring or early summer of 1962, before any of Kennedy's warnings.[98] Commitments make policy makers less receptive to warnings.

By 4 September, when Kennedy issued his first specific warning, the Soviets had already embarked on a clear course of action. They had secured Castro's approval for the missiles, they had shipped men, building materials, and weapons to Cuba, and they were well advanced with the construction of the actual missile sites. To have backed down at this point would almost certainly have endangered relations with Castro. If word of the fiasco leaked out, as it almost certainly would have done, the Soviet Union would also have lost face with China, which had repeatedly charged Moscow with undue timidity. At home, withdrawal could have caused a serious rift within the Soviet leadership. Khrushchev would have been pilloried by his enemies for embarking upon a harebrained scheme and then losing his nerve in the course of its execution. Indeed, Khrushchev could have concluded that backing down in September would give his enemies the opening they needed to topple him from power. But he must also have realized that, if Kennedy meant what he said, to persevere held out the prospect of a serious confrontation with the United States.

This is precisely the kind of decisional conflict that encourages defensive avoidance. Policy makers who perceive that serious risks are inherent in their current policies, but upon reflection are unable to identify an acceptable alternative, experience psychological stress. They become emotionally aroused and preoccupied with finding a less risky but nevertheless feasible policy alternative. If, after further investigation, they conclude that it is unrealistic to hope for a better strategy, they will terminate their search for one even though their dissatisfaction with the available option continues. They will also become insensitive to warnings that indicate the course of action to which they are now committed will not succeed.

Soviet leaders, being no more and no less human than the rest of us, may well have responded to their decisional dilemma by bolstering the policy to which they were committed. If so, they could easily have exaggerated to themselves the positive consequences of successfully deploying missiles in Cuba. At the same time they would have minimized their assessment of any serious American reaction and would have made every effort to reinterpret or discredit information that suggested otherwise. In this regard Ambassador Anatoly Dobrynin, known by reputation as a man reluctant to challenge the prevailing orthodoxy in Moscow, may have helped them by watering down the impact of the private warnings he received from members of the Kennedy administration so as not to offend his political superiors. Soviet leaders could therefore have maintained their comforting illusions until reality rudely intervened, in the form of Kennedy's announcement of the blockade. Fortunately for everyone, they possessed the psychological resources to recognize their miscalculation and back away from their initiative.

The View from Washington

American officials also struggled to find an effective policy to safeguard their interests and tried to predict how their adversary would respond. They did not miscalculate the way the Soviets did. But in my opinion they came very close to doing so.

Opinion in the "Ex Com" swung back and forth between an airstrike and a blockade before a consensus finally developed in favor of the latter. Even so, a minority of Kennedy's advisers remained convinced that the only effective means of putting an end to the threat posed by the Soviet missiles was an airstrike. Kennedy himself was attracted to the airstrike for this reason. He decided against it only after the air force had twice demurred from guaranteeing that it would destroy all of the missiles.

The president, to his credit, was also concerned about the possible Soviet reaction to the airstrike, because it was certain to kill Russians in the vicinity of the missile, radar, or anti-aircraft missile sites. Would Khrushchev be compelled to respond with a military initiative of his own, he wondered, perhaps against the American missiles in Turkey, setting in motion a chain of escalating reprisals that could end in nuclear war?[99] After the crisis, Kennedy told the American people: "Above all, while defending our own vital interests, nuclear powers must avert those confrontations which bring an adversary to the choice of either a humiliating defeat or a nuclear war. To adopt that kind of course in the nuclear age would be evidence only of the bankruptcy of our policy—or of a collective death wish for the world."[100]

Many of Kennedy's advisers were remarkably sanguine about the risks of an airstrike. Dean Acheson, Paul Nitze, John McCone, Douglas Dillon, and the Joint Chiefs were all convinced that the missiles should be attacked and that the Russians would do nothing in response. They argued the case for the airstrike during the Ex Com's deliberations, and for much of the time they represented the majority position. Indeed, Robert Kennedy reported that at least one high-ranking military official urged the president to order airstrikes against Cuba *after* the Soviets had capitulated![101]

High-ranking air force officers also regretted that the United

States had failed to exploit its military superiority. Four months after the crisis one of McNamara's civilian analysts listened to Curtis LeMay, then head of the Air Staff, address a group of air force officers on the lessons of Cuba. "The Soviets are rational people," LeMay insisted. American nuclear superiority and conventional advantages in the Caribbean guaranteed that there had been "no real risk" of war during the crisis. "The problem," LeMay declared, "had been the flap at the White House." "The thing to do next time was to head these people off."[102]

Such swaggering self-assurance was not limited to LeMay and "superhawks" like him. It also affected more responsible political officials. Dean Acheson felt so strongly the United States should have gone ahead with the airstrike that he subsequently publicized his opposition to Kennedy's policy. In a magazine article he charged that Kennedy had triumphed only because of "plain dumb luck."[103] Acheson's judgement was echoed in the official postmortem on the crisis, prepared by Walt Rostow and Paul Nitze in February 1963. They concluded that the principal error of the president and his advisers was that they had worried too much about the danger of nuclear war: "They argued that this exaggerated concern had prompted consideration of improvident actions and counseled hesitations where none were necessary. Since the United States could get its way without invoking nuclear weapons, the burden of choice rested entirely on the Soviets."[104]

The first scholarly analyses also took the position that Kennedy had exaggerated the likelihood of war.[105] The Wohlstetters' study, the most highly regarded of these analyses, argued that the decisive factor had been American military superiority at every stage of escalation. "Chairman Khrushchev," the authors asserted, "stepped down to avoid a clash of conventional forces in which he would have lost." Escalation to the nuclear level would have only made matters worse. American nuclear capability "would have made a Soviet missile strike against the United States catastrophic for Russia."[106]

A quarter century of hindsight have not altered the conviction of many airstrike advocates that the United States missed a golden opportunity to topple the Castro regime and teach the Soviets a much-needed lesson. Their conviction, then and now, derives entirely from their assessment of the balance of military power. The balance unquestionably favored the United States in the case of either a conventional war in and around Cuba or a strategic nuclear exchange. Analysis of this kind nevertheless ignores all the other considerations that could have influenced the Soviet response to an airstrike. Certainly Soviet leaders would have taken into account their estimates of the foreign policy costs of a crushing military defeat, the possibility that acquiescence to this kind of humiliation would encourage further American aggression against them, and, of course, the likely domestic repercussions for Khrushchev and his supporters in the Kremlin. In short, some of the very same reasons that made Kennedy ready to risk war when he discovered Soviet missiles in Cuba could have compelled Khrushchev to do the same in response to an American airstrike.

No one really knows how the Soviet Union would have reacted, but Soviet leaders would certainly have consulted their political interests as well as their military capabilities. It is conceivable that Khrushchev would have bowed to American military might, but it is also possible that he would have escalated the crisis by responding with some initiative of his own, in the hope of forcing a compromise settlement. History is full of examples of states that took up arms against militarily superior adversaries for fear of the costs of not doing so. Finland in 1940, China in 1950, and Egypt in 1973 are cases in point. Leaders of all three countries went to war because they believed that war was the only way to defend vital national interests. That American policy makers could smugly conclude that the Soviet Union would tolerate an airstrike against its missiles, airbases, and five thousand military personnel in Cuba is nothing short of remarkable. For them to have based their analysis purely on the military balance is more incredible still.[107]

The perspective of a quarter century makes it obvious that Ex Com participants, regardless of what option they favored, had little insight into the motivations of their adversary. Almost without exception, they approached the analysis of Soviet foreign policy in relatively simplistic and largely ideological terms. Ambassador Charles Bohlen, the Ex Com's Soviet expert, explained Khrushchev's action as a probe of American resolve. He quoted, apparently to great effect, Lenin's admonition to his fellow Bolsheviks that foreign policy was like a bayonet thrust: "If you strike something hard, pull back, but if you hit mush, keep on going."[108] Bohlen's characterization of Soviet policy was almost universally accepted by Ex Com participants, including the president himself.

Because Khrushchev's behavior was interpreted as pure adventurism, he could and would back down, it followed, when confronted by superior force and resolve. Indeed, Khrushchev had to be met with as much firmness as possible in order to discourage future challenges.[109]

Such a cardboard characterization of the adversary made empathy all but impossible. Ex Com members could not imagine any important Soviet interests that might have motivated Khrushchev's reckless initiative. They never considered the hypothesis, now accepted by most students of the crisis, that Khrushchev's furtive ploy was primarily a reaction to domestic fears of the political consequence of the Soviets' own strategic weakness. Blind to the interests that the Soviets might have believed they had at stake, so many members of the Ex Com were also blind to the dangers of pushing the Soviets too far. They failed to see what Khrushchev had to lose, public face aside, by backing down. Some even denied the importance of face, arguing that Khrushchev had no domestic public opinion with which to contend.

It is to Kennedy's credit that he was able to transcend this simplistic idea of his opponent in order to work out something of a compromise settlement. His diplomacy in the later stages of the crisis was characterized by flexibility and sensitivity. It is nevertheless true that, given the premises that guided American policy in the confrontation, the president—and Khrushchev too—did indeed benefit from some "plain dumb luck," albeit for reasons quite different from those suggested by Acheson.

Kennedy's tactical skill contrasted sharply with his strategic blundering. His administration's inability to grasp Moscow's concern for its pronounced strategic vulnerability helped precipitate the Cuban crisis and intensify dramatically the arms race. The "lessons" the White House drew from that confrontation acted as a catalyst for subsequent American military intervention in Vietnam. There is little reason to suppose that contemporary American policy makers have any better understanding of what motivates Soviet leaders. Reagan, Weinberger, and Schultz routinely describe Soviet policy in simplistic ideological terms. Carter and Brzezinski, though less confrontatory in their rhetoric, were no more successful in predicting or explaining Soviet foreign policy. The same has been true for their professional advisers. The Soviet intervention in Afghanistan took the intelligence community by surprise, for example, despite the fact that the prior military buildup was successfully monitored. The foreign policy "experts" then proceeded erroneously to predict a Soviet invasion of Poland in December 1980. They still have little grasp of the reasons that prompt the Soviets to use force or how the Soviets assess the risks associated with such ventures.

Many analysts, in and out of government, at first sought to explain Soviet-Cuban military involvement in Africa and the Soviet intervention in Afghanistan as the results of a change in the "correlation of forces" which favored the Soviets. Acting under the umbrella of nuclear parity and conventional military superiority, the argument goes, Moscow is no longer deterred by the West and is on the lookout for opportunities to expand its sphere of influence.[110] More sophisticated interpretations exist, to be sure, but their proponents never had the ear of the Reagan administration. Nor, to be fair, did they represent the dominant voice among its Democratic opponents.

Analysis based primarily on calculations of relative military advantage does not augur well for American success in predicting or responding effectively to future Soviet foreign policy initiatives. Indeed, it could prompt more confrontatory behavior in crises in which Americans believed they held the advantage.[111] Such a policy, if the Soviets had important interests at stake, could increase the prospect of war by miscalculated escalation.

Coping with Crisis

Efforts to minimize the probability of miscalculated escalation must address the two fundamental causes of the phenomenon. Policy makers must be encouraged both to develop more empathy for their adversaries and to learn more about the details of crisis management and war plans. Empathy is the more important of the two, of course, because it is necessary not only to manage crises but to avoid them. Empathy requires policy makers and their advisers to develop more insight into the fears and aspirations of their adversaries, the kinds of pressures and constraints that act upon them, and the organizational context in which they must make and implement decisions.

Lack of empathy encourages two kinds of interpretative distortions. The first is the belief that one's adversary is implacably hostile. This is rarely true; Nazi Germany is the exception, not the rule. Most international conflicts combine hostility, misunderstanding, and clashes of interest. Usually even the most acute rivals have some common interests. Unfortunately an image of the adversary as an "evil empire" tends to rule out all possibility of constructive dialogue or negotiations. It makes policy makers unreceptive to overtures or pacific gestures because it predisposes them to put the worst possible interpretation on any statement or action by the other side. The history of the Cold War is full of opportunities missed because of the wall of mutual cognitive mistrust which divides the superpowers.

Lack of empathy also blinds policy makers to the interests, pressures, and constraints that shape their adversary's decisions. This blindness may be a special case of what psychologists call the fundamental attribution error, a cognitive bias that inclines people to see the actions of others as expressions of basic predispositions while they see essentially the same actions on their own part as responses to situational pressures.[112] Or it may be the result of motivated biases of the kind responsible for Austrian miscalculation in 1914 and Soviet miscalculation in Cuba. Either way, the effect is the same. Policy makers convince themselves that their adversary will tolerate a provocation because its leaders understand all the pressures compelling you to act. As they are not thought to be subject to any of these pressures themselves, it is also assumed that the adversary will have no need to respond in kind.

Correcting faulty judgments of this kind is a formidable task. Sometimes cognitive biases can be overcome by education.[113] More sophisticated policy makers, alert to the pitfalls of oversimplified and tautological information processing, are less likely to make such mistakes. Images of the adversary as implacably hostile and ruthlessly immoral are, however, an integral part, perhaps even a defining characteristic, of conflictual relationships. Urging officials to take a more benign, or at least nuanced, view of their adversary is the political equivalent of tilting at windmills. Even those policy makers who are capable of grasping the complexity of important foreign policy issues will still have difficulty in understanding an adversary's behavior if their analysis rests, as it so often does, on stereotyped notions of its leaders' motives.

Even more imposing difficulties stand in the way of overcoming motivated bias. Little can be done to educate policy makers who, like John Foster Dulles, have deep-seated personal needs to conceive of the world as a Manichean struggle between good and evil. Even more open-minded leaders are likely to shut their minds to reality when they are on the horns of a serious decisional dilemma. Policy makers who feel compelled to pursue a specific foreign policy tend to become insensitive to the interests and commitments of other states that stand in the way of the success of that policy. Numerous twentieth-century crises and wars attest to the remarkable obtuseness of leaders presented with warnings that they are headed toward disaster. To influence policy makers in such situations, it is necessary to ease the kinds of pressures that are acting on them or somehow to convince them that the interests they are trying so desperately to preserve can be protected at least as well by a different policy.

Ignorance of war plans and lack of interest in crisis management is a problem more discrete than lack of empathy. At first glance it also appears more amenable to treatment; policy makers must simply be made to face up to their responsibilities. Nevertheless there are important practical and psychological barriers standing in the way of this goal. Chief among these is the propensity of national leaders to practice defensive avoidance as a means of coping with the anxiety associated with war plans and crisis management.

Defensive avoidance is made easier by the burden of office, which compels a president to make very selective and restrictive choices about how to allot working hours. In the rush of business, responsibility for crisis management and war planning can be allowed to slip through the cracks. It is made easier still by the fact that most presidential advisers have no wish to confront such anxiety-provoking problems. They are unlikely to do anything to encourage a president to face the issues. In this conspiracy of silence the chief executive goes about the nation's business more or less insulated from any thoughts about nuclear crisis management. The only ugly reminder is the omnipresent officer with the "football"—the satchel with the go-codes. It would not be at all surprising, however, to learn that most presidents find some way of insulating themselves from even this palpable symbol of their responsibilities.

This presidential reluctance to consider the possibility of a nuclear crisis or nuclear war is by no means altogether irrational behavior. Presidents must cope with so many domestic and foreign crises that it would make little sense for them to devote many hours of an already severely cramped schedule to problems that may never happen. Once they make such a decision, presidents will almost inevitably reduce in their own minds even further the probability that these problems will ever require their serious attention. By doing so, they enhance their own peace of mind and with it their ability to function. Statistics also make defensive avoidance seem politically sensible, for the likelihood that a given president will actually have to manage an acute nuclear crisis is low. The last such crisis was in 1962, fully a generation ago. Since then, five presidents have held office.[114]

Although the odds are against any president having to manage a nuclear crisis of the magnitude of Cuba, it would be foolish to assume that *no* future president will ever have to do so. What is in effect an efficacious psychological strategy for an individual president could prove catastrophic to the national interest if a succession of presidents adopt it. For if a nuclear crisis arises, the president's ability to cope with it is likely to be inversely proportional to his or her previous practice of defensive avoidance. This is true for several reasons. The first pertains to experience: successful crisis management requires knowledge that cannot readily be assimilated in the course of a confrontation. Presidents who have not previously involved themselves in the details of crisis management are not likely to be sufficiently aware of the danger of loss of control which is associated with high alert levels. They are also more likely to become captives of prepackaged military options that bear little relationship to their political needs at the time. Finally, they are more likely to be affected adversely by the stress of the crisis.

Stress has both external and internal causes. Someone given a task that is too complex to perform in the allotted time experiences stress. So does a person confronted with information or impulses that threaten important goals or his or her personality structure. In a crisis these two kinds of stress may prove to be mutually reinforcing. Political leaders and their advisers are likely to work long hours under great tension, trying to manage a situation they fear may be beyond their abilities. They are also likely to worry that an oversight or false move on their part could result in catastrophe. If previously they have denied even the prospect of such a confrontation, they may lack the skills or knowledge they need to cope. Their self-confidence may also be shaky or nonexistent. In these circumstances they will feel stress more acutely, and their performance will likely decline.

Clinical studies reveal that a person suffering from acute stress displays apprehension, helplessness, and a pervasive sense of nervous tension.[115] Other clinical manifestations may include headaches, indigestion, anorexia, palpitations, genitourinary problems, insomnia, irritability, and inability to concentrate. When confronted with the need to act, highly stressed people are likely to be indecisive. They may oscillate between opposing courses of action, unable to accept responsibility for a decision. They are also likely to mistrust their own judgment and be easily influenced by the views of others. Their ability to perform tasks effectively when those tasks relate to the source of anxiety will be low.[116]

The problem of stress has obvious and profound implications for crisis decision making at the individual, group, and organizational levels. Case studies illustrate some of the ways in which stress degrades the performance of cognitive and judgmental tasks. Stress also was at least partially responsible for bad decisions that led to unanticipated and undesired wars.[117]

Drawing upon the work of others, especially that of Ole Holsti, Alexander George has formulated a list of ways in which severe crisis-induced stress can impair crisis management. The findings that he synthesizes indicate that severe stress can affect almost every important aspect of the decision-making process.

Impaired Attention and Perception

—Important aspects of the crisis situation may escape scrutiny.
—Conflicting values and interests at stake may be overlooked.
—The range of perceived alternatives is likely to narrow but not necessarily to the best option.
—The search for relevant options tends to be dominated by past experience; there is a tendency to fall back upon familiar solutions that have worked in the past whether or not they are appropriate to the present situation.

Increased Cognitive Rigidity

—Impaired ability to improvise; reduced creativity.
—Reduced receptivity to information that challenges existing beliefs.
—Increased stereotyped thinking.

—Reduced tolerance for ambiguity leading to a cut-off of information search and premature decision.

Shortened and Narrowed Perspective

—Less attention is paid to longer-range considerations and to the consequences of contemplated actions.
—Less attention is paid to the side-effects of options.

Shifting the Burden to Opponents

—A belief that one's own options are quite limited.
—A belief that the opponent has it within his power to prevent an impending disaster.[118]

Stress is most damaging when it leads to wildly erratic behavior or temporary incapacitation. In extreme circumstances it may result in a "dissociative reaction"—a mental state in which a person's thoughts, feelings, or actions are not associated or integrated with important information from the environment. Simply put, a dissociative reaction is a jamming of one's circuits. Clinical symptoms include trance states (characterized by unresponsiveness to the environment, immobility, and apparent absorption with something deep within the self), estrangements and paramnesia (detachment and disengagement from persons, places, situations, and concepts), fugue (flight, entered into abruptly, often with amnesia and lack of care for one's person or surroundings), frenzied behavior (episodes of violent, outlandish, or bizarre behavior), and dissociative delirium (including hallucinations, wild emotional outpourings, and the release of primary process material).[119]

Recent history reveals that dissociative reactions are by no means a rare occurrence among political leaders. In 1914, at the height of the July crisis, Kaiser Wilhelm suffered a temporary psychiatric incapacitation, commonly referred to as a nervous breakdown. He remained incommunicado for twenty-four hours at his residence in Potsdam. In June 1941 Stalin had a breakdown after learning that Germany had invaded the Soviet Union. He was out of touch with events for several days. Eden in 1956, Nehru in 1962, Rabin and Nasser in 1967 were also incapacitated, mentally and physically, during the course of a prewar crisis or in the aftermath of national setbacks.[120] Nasser, for example, withdrew to the solitude of his villa. According to Anwar Sadat, "he looked at the time—and for a long time afterwards—[like] a living corpse. The power of death was evident on his face and hands although he still moved and walked, listened and talked."[121]

The German and Russian cases are particularly germane to the strategic dilemma that has concerned us throughout this book: the sometimes contradictory policies required to reduce the risk of war and guarantee retaliation. Fright, rage, depression, or more extreme responses to stress can interfere with both goals. In Germany these emotional responses were a major reason why diplomacy's failure to avert war. In the Soviet Union, Stalin's defensive avoidance prevented adequate defensive preparations on the eve of the German invasion.

At the height of the July crisis, German kaiser and chancellor alike displayed irresolution, bewilderment, and loss of self-confidence. The kaiser oscillated between profound optimism and despair, and he suffered a temporary breakdown on 30 July. His hypervigilant chancellor swung back and forth between the very extremes of available policy options. At first he favored decisive action by Austria. Later, appalled by the thought of a European war, he sought to restrain Vienna. Influenced by Moltke, he again reversed himself, urging Austria to declare war on Serbia. Both kaiser and chancellor ultimately lasped into passive acceptance of the inevitability of war, although they continued to clutch at the hope of British neutrality the way drowning men grope for a life preserver.[122]

The near paralysis of the German political leadership amounted to de facto abdication of control over the course of policy during the most fateful hours of the crisis. This breakdown in political decision making meant not only that German policy continued on a predestined course but that the military, pushing for mobilization, met only half-hearted resistance from political leaders. The general staff effectively made the decisions that plunged Europe into a catastrophic war that hardly anyone else desired and that probably could have been avoided.

The German experience highlights the need for governments to learn from their past behavior and to modify their policies in response. Acute international crises place a special premium on this ability, because crisis strategies are based on a set of expectations about adversarial behavior. If these expectations are wrong, policy must be revised. And it must be revised with dispatch, because crisis decision making is generally subject to severe time constraints. As learning and steering capability diminish, policy comes to resemble a stone rolling downhill; it cannot be recalled, nor can its path be altered. If the assumptions underlying policy are incorrect, as was true of Germany in 1914, the crisis may well result in war despite leaders' desires to the contrary.

Stalin's breakdown had similar causes. The Soviet dictator had convinced himself that what he feared most of all—a German invasion of the Soviet Union—would not happen. Like Wilhelm, he resorted to denial and paranoid projection to cope with the evidence accumulating to the contrary. He dismissed repeated warnings of invasion from both western and Soviet officials as calculated attempts to embroil Russia in a conflict with Germany.[123]

When the German onslaught shattered his illusions, Stalin became temporarily paralyzed. Ivan Maisky later revealed that "from the moment of the attack by Germany, Stalin locked himself in his study, would not see anybody, and did not take part in state decisions."[124] Marshal Zhukov remembered that when he telephoned Stalin and told him of the German attack, a long silence followed, punctuated only by the Soviet leader's loud breathing. Zhukov asked if Stalin had understood him. After a prolonged silence, the voice on the other end of the line told him to have Poskrobyshev, Stalin's secretary, summon the Politburo. At this meeting, attended by generals Zhukov and Timoshenko, Stalin sat pale and silent, cradling his unlit pipe, his mind apparently somewhere else.[125]

Stalin's withdrawal brought Soviet policy to a standstill. For Stalin, as Admiral Kuznetsov observed, so dominated the men around him that when the crisis came, "they could not take in their hands the levers of direction."[126] This indecision cost the Soviets dearly in lives and territory in their struggle against Germany. The initial Soviet war effort was also hindered by Stalin's belief, even after he had received Germany's formal declaration of war, that the invasion was an unauthorized provocation on the part of the German generals. In keeping with this view, Stalin ordered the Soviet Air Force not to penetrate more than ninety miles beyond the frontier, kept open the radio link with the German foreign ministry, and asked Japan to mediate the conflict. For some months afterward he continued to harbor the idea of a negotiated peace based on Russian concessions in eastern Europe.[127]

Recent Soviet and American leaders appear to have done their best to convince themselves that they will never confront a nuclear war. To the extent that, individually and collectively, they doubt their ability to cope with such a threat, they would be extremely reluctant to give credence to warnings that their country was about to be attacked. Like Stalin, they could go to great extremes to deny the ugly reality, even acting in ways that increased their country's vulnerability.

Confronted with unambiguous evidence of nuclear attack, a Soviet premier or an American president could seek refuge in a trance state, paralysis, or some other kind of extreme defensive reaction in order to protect his ego from information that was too threatening to confront on a conscious level. What happened next would depend upon the arrangements that had been made to delegate launch authority or provide for someone to step into the breach if the leader became incapacitated. But a breakdown of this kind could seriously interfere with retaliation, perhaps prevent it altogether.

The stress generated by a nuclear crisis would almost certainly surpass that of the most acute crises of the past. Even the healthiest political and military leaders have strong incentives to deny the likelihood of such an event. Indeed, what little evidence we possess suggests that this is just what happens. The combination of the incredible stress of a nuclear crisis *and* the efforts of leaders beforehand to deny the likelihood of such an event seems a sure recipe for psychological disaster. A nuclear crisis would almost certainly produce some kind of unexpected and dysfunctional behavior. Cuba certainly did. In 1984 one high official in the Kennedy administration revealed that two important members of the Ex Com had been unable to cope with the stress of that confrontation; they became entirely passive and were unable to fulfill their responsibilities.[128]

Even if a future crisis did not result in anything as dramatic as a leader's breakdown, slow or inadequate performance of key individuals in either country's chain of command could still produce a partial paralysis. Behavior of this kind could impede the prospects either of crisis resolution or of retaliation. To date, this problem has attracted little attention from those who study nuclear strategy.

Coping with Stress

The possibility that stress could prompt dysfunctional behavior that would cause or contribute to miscalculated escalation requires us to look for ways of minimizing its presence and effects. Social psychologists, albeit in quite different contexts, have studied the causes and effects of stress and possible means of coping with it. They have devised techniques that show some success in counteracting defensive avoidance, promoting vigilant decision making among people who face stressful dilemmas concerning their health, careers, and other personal issues. The utility of these techniques for national leaders dealing with policy issues has not yet been investigated.

"Awareness of Rationalizations." This technique is designed to reduce defensiveness about information that indicates the risks of continued denial. The individual is required to recognize the rationalizations he or she routinely employs to resist information that would reactivate stress aroused by an internal conflict. Once the individual comes to understand the role these rationalizations play, he or she is more likely to engage in an honest exploration and frank acknowledgment of basic, deep-seated thoughts and feelings.[129]

Emotional Role Playing. In a realistic psychodrama, individuals act out their fears in order to come to terms with their own vulnerability. Role playing is often successful in undermining the common cognitive defense: "It can't happen to me." It has been used in smoking clinics, for example, to reduce cigarette consumption by making smokers acknowledge their personal vulnerability to lung disease. Modifications of the technique are now being developed for application to other contexts, including policy decisions.[130]

Stress Inoculation. A procedure designed to produce positive behavioral change by enhancing an individual's sense of self-confidence, stress inoculation has for years been an integral part of basic training for combat soldiers and pilots, both civilian and military. By providing a repertoire of skills useful in emergencies, this technique can help reduce the disruptive effects of fear in two ways. It reduces the general level of anxiety people feel when contemplating threatening situations. It also reduces the intensity of fear reactions in actual emergencies once people begin to respond in a skilled manner.[131]

The Balance Sheet Procedure. This is a predecision exercise that requires an individual to confront and answer questions about potential costs, gains, and risks that have not previously been considered. Such an exercise is particularly valuable in a situation where the person has, for whatever reason, avoided giving serious thought to the problem and possible responses to it. When successful, the procedure results in increased vigilance, self-disclosure, stress inoculation, and self-persuasion.[132]

Outcome Psychodrama. This technique seeks to induce individuals to explore and evaluate more fully the consequences of their behavior or the alternatives they prefer. People are asked to project themselves into the future and to assess the worst possible consequences of their behavior or decision. By doing so, they may become aware of expectations, attitudes, and outcomes they previously avoid thinking about. In some circumstances, however, psychodrama may actually intensify defensive avoidance.[133]

We must approach all of these techniques with great caution. They were pioneered in quite different contexts; surgical wards, smoking clinics, and pilot-training courses are a far cry from the National Command Authority in the midst of a nuclear alert. The stress that national leaders would confront in such a situation would almost certainly be several orders of magnitude greater than what a smoker feels when contemplating personal vulnerability to lung cancer. But the stress experienced by a patient about to undergo a life-or-death surgical procedure is probably every bit as extreme, even though only one life is at risk. As a result, some of the lessons psychologists have learned are likely to be germane to crisis management.

Several of the techniques I have noted encourage people to develop self-confidence in their ability to cope successfully with a future threat or a problem at hand, or they seek to reassure people by convincing them that their fears are exaggerated. Clearly neither is appropriate in the case of a nuclear crisis; the task before us is to convince policy makers just how difficult a crisis involving nuclear alerts would be to control, not to reassure them that everything will turn out all right. Such techniques, however, might be relevant for coping with lower-order crises of the type that could, if mismanaged, lead to more acute and dangerous confrontations.

Another point to consider is that in most studies of psychological intervention, the smokers, surgical patients, or would-be pilots were volunteers. Many of them were motivated by a desire to change their behavior, and that desire could itself be an element essential to the success of any of the procedures we have examined. Smokers coerced into attending a smoking clinic are not good candidates for psychological intervention.[134] Yet their situation is analogous to the one we confront with policy makers. To date, no officials have volunteered to attend a "crisis management clinic"—nor are they likely to if left to their own devices. Some means must be found of inducing—not coercing—them to avail themselves of the various techniques that might be helpful in coping with a nuclear crisis.

Finally, we must recognize that the success or failure of all of these procedures is measured in statistical terms. Even with a demonstrably high success rate, a given procedure may not work in the case of a specific individual. When we talk of nuclear crisis management, of course, we are primarily concerned with a small group of individuals: the president and his immediate advisers. For this reason, the success or failure of a procedure will depend very much on the personalities of the leaders in question. We must therefore pay as much attention to idiosyncratic as to situational variables when we formulate strategies for inducing national leaders to take a more active role in crisis management planning.[135]

One possible mechanism to force presidents to confront the problems associated with crisis management and nuclear war is legislation requiring periodic briefings or command post exercises. The SIOP briefing a new president currently receives is quite perfunctory and superficial. It provides him with some essential information about the nature of American war plans, but it does not address nuclear alert procedures or the problems of crisis management, even though this kind of knowledge is essential to a president. One possible drawback to congressional action, assuming for the moment that it is politically feasible, should be noted, however: it could be perceived by presidents and their advisers as coercive. If so, new legislation along these lines would probably be self-defeating.

A better idea might be a blue-ribbon panel representative of different political points of view and composed of psychologists, former political and military officials, and political scientists. Recommendations of the kind we have discussed coming from such a panel, followed up perhaps by personal interviews with a president or president-elect by some of its members, might be far more effective in convincing the president of his or her responsibility. A willing president, and advisers, could then be exposed more profitably to briefings and exercises that incorporate some of the techniques I have mentioned.

Is it possible that familiarity with American war plans and crisis management procedures could make the president and his advisers less frightened by the prospect of nuclear war, or unrealistically confident about their ability to manage a nuclear crisis? Such a prospect cannot categorically be ruled out, even though it seems remote. The facts of the case certainly do not encourage sang-froid or confidence. My own experience with briefings of this kind, for what it is worth, is that officials exposed to them almost always come away with a more sober appreciation of the problem. The same would probably also be true for a president.

Greater familiarity with crisis management procedures and war plans would make a president more aware of just how difficult to manage a serious superpower crisis would be. The knowledge would be likely to make the president more rather than less cautious. Periodic contact with war plans, alert procedures, and crisis management issues would also make it more difficult for a president to practice defensive avoidance. Both developments might provide a strong incentive for the president to do something to reduce the chance of such a confrontation coming about. Of course, there is always the possibility that a president would seek refuge in some other form of psychological escape. This may have happened with Ronald Reagan, whose SIOP briefings seem to have been the catalyst for his commitment to Star Wars.

Overcoming presidential reluctance to contemplate the possibility of a nuclear confrontation would remove an important impediment in the way of crisis management. But it would not guarantee that the president or top officials would function effectively in a crisis. The stress would still be enormous, and it could seriously affect the performance of even the best-briefed and most knowledgeable leader. And as I have argued, this stress would be much greater still if the policy makers in question had previously resorted to denial to downplay the likelihood of a crisis.

The primary precipitating factor of the several breakdowns of political leaders I have noted was not the external stress of the crisis. Rather, it was the shock of adversaries acting in ways they had denied to themselves they would. In all these situations the personality or political needs of leaders conflicted with external realities; but leaders chose to indulge the former at the expense of the latter. The kaiser convinced himself that Russia, France, and Britain would not intervene in a Balkan war; Stalin, that Germany would not invade the Soviet Union; and Nehru, that China would never use force against India. All three leaders did their best to insulate themselves from information that challenged their unrealistic expectations. Their incapacitation came after events shattered their illusions. It was an extreme means of defending their egos against material that was too threatening for them to cope with.

These several cases highlight the relationship between the origins of crisis and performance during a crisis itself.[136] When wishful thinking, self-delusion, and defensive avoidance undergird the foreign policies that bring a crisis about, crisis management will suffer. Because leaders have done so much to insulate themselves from threatening information, they are likely to realize only belatedly the extent of their miscalculations. By that point they may be unable to set the situation right, either because of their own loss of self-confidence or because the crisis has progressed too far along the path to war.

We have come full circle. I began this book with the assertion that crisis stability has structural causes; that neither technical nor procedural improvements, welcome as they would be, could succeed in bringing about crisis stability. Instead, we have to do something about the strategic, political, and psychological realities that trigger crises and shape their outcome.

The vulnerability of command and control is one such reality, as the chapter on preemption made apparent. It is a major cause of crisis instability, and one that we can do very little about in the absence of some far-reaching (and unlikely) breakthrough in arms control. The chapter on loss of control brought home the same point from a different perspective. It outlined the ways in which efforts to compensate for command and control vulnerabilities have heightened the prospects of accidental or inadvertent war—another fundamental and enduring cause of crisis instability. Finally, the current chapter has explored the structural impediments to good decision making. It has tried to show how those impediments derive from the very nature of nuclear crises and their origins. These findings indicate that crisis instability is a problem more intractable than many authorities suppose. They also point to the need for initiatives different in kind from those usually envisaged as means of addressing the problem.

Conclusions

[5]

Toward Crisis Stability

This book has used the origins of World War I as a vehicle to explore the political-military conditions that affect crisis management. Many of the most important causes of war in 1914, I have argued, are also salient features of contemporary international relations. They could again hinder the resolution of an acute crisis. This said, however, it is important to recognize major differences between the two historical periods which make war less likely today than it was in 1914. The most important of these pertains to expectations about the inevitability of war.

Perceptions of War: 1914 and Today

By 1914 opposing alliances, arms races, and a series of diplomatic confrontations had engendered a pervasive sense of pessimism in European capitals. In Vienna the idea of a preventive war to save the empire was openly discussed; it had strong advocates in both the army and the Foreign Office. Austria's adversary, Serbia, prepared for war in the hope that it would serve as midwife to a South Slav national state. In St. Petersburg the political and military elite also anticipated war. Anti-German feeling in Russia, especially among the czar's entourage, had increased markedly after the Balkan wars. There were many who welcomed the prospect of a conflict as an opportunity to settle old scores. In Rome nationalists and irredentists, but also some political conservatives, saw war as unavoidable and desirable. Even in London and Paris successive European crises, which aroused fear and concern about

German objectives, had brought many to believe that war in the not too distant future was a real possibility.[1]

It was in Berlin that the expectation of war was most pronounced. Wolfgang Mommsen observes that "wide circles of the German public, as well as thier political leaders, awaited the approaching war with an alarming spirit of fatalism."[2] This attitude derived in large part from German perceptions of their encirclement by Russia, France, and Britain. German leaders had sought to break out, but their clumsy efforts had succeeded only in bringing the three powers closer together. Many Germans had become positively paranoid about the dangers they alleged they faced. In 1909 no less a personage than Field Marshal Alfred von Schlieffen expressed the fear that

> an endeavor is afoot to bring all these powers [Russia, the Balkan states, Italy, France, and Britain] together for a concentrated attack on the Central Powers. At the given moment the drawbridges are to be let down, the doors are to be opened and the million strong armies let loose, ravaging and destroying. Across the Vosges, the Meuse, the Niemen, the Bug and even the Isonzo and the Tyrolean Alps. The danger seems gigantic.[3]

If war was only a matter of time, it followed, in the words of German foreign minister Gottlieb von Jagow, that "one must not allow the enemy to dictate the moment but determine it oneself."[4] From the general staff's point of view, the dictum meant a decision to go to war sooner rather than later, because the generals were convinced that time was working against Germany. Both France and Russia were known to be improving their military capabilities vis-à-vis the Reich. The generals were particularly concerned about Russia, then in the process of building a strategic railway network, modernizing armaments, and expanding the size of its army by 40 percent. These reforms, financed by French credits, were expected to reach fruition by 1917 and render the Schlieffen Plan obsolete.[5]

Chief of Staff Helmuth von Moltke drew the obvious conclusion. In December 1912 he told the kaiser: "The sooner there is war the better."[6] Although his advice was not followed at the time, Moltke continued to push for a decisive confrontation—as he had been doing since 1909. In the spring of 1914 he went so far as to ask the government to start a war as soon as possible, because every moment Germany hesitated diminished the advantage. The foreign minister and chancellor rejected Moltke's plea, but both later admitted that the chief of staff's insistence that war not be postponed influenced their behavior during the July crisis.[7]

The idea of preventive war clearly obsessed Moltke, although the concept also found widespread support in the foreign office and the Chancellery. So it was that the chief of staff and those around him saw in Sarajevo the opportunity for Germany to draw the sword. Throughout the crisis both Moltke and Count Waldersee, his quartermaster general, urged their political counterparts to act lest they lose what might prove to be their last opportunity to escape encirclement. Moltke was so fearful that Wilhelm would back down yet again that he worked behind the back of both the kaiser and the chancellor, urging Austria to go to war without delay.[8] The chief of staff was not to be denied his preventive war.

Neither the United States nor the Soviet Union (with the possible exception of the Cuba crisis), as far as we can tell, has ever felt the same desperation about the strategic balance. Each superpower fears the other's nuclear arsenal, of course, and the consequences of strategic inferiority. In the 1950s and 1960s the Soviet Union was clearly at a disadvantage because of its limited means of striking at the United States. In the late 1970s and the early 1980s many American strategists spoke with some foreboding of the "window of vulnerability" they believed the Soviet Union had opened up by virtue of its impressive counterforce capability. In the mid-1980s it is the Soviet Union that is feeling increasingly threatened, because of the array of new American weapons now being deployed.

Seesawing perceptions of vulnerability have certainly affected the policies of the superpowers, but they have never made them desperate in the way German leaders were in 1914. Indeed, in only one instance can a convincing argument be made that fear of strategic disadvantage pushed a superpower into a truly risky initiative. This was the Cuban missile crisis, touched off by Khrushchev's decision to put missiles furtively into Cuba. Many Western analysts believe his initiative was primarily a response to perceptions of acute strategic threat.[9]

The absence of strategic pressures of the kind Germany faced in 1914—which made war seem the only way to preserve German security—differentiates present international relations from the situation that prevailed on the eve of World War I. It is one reason why peace between the superpowers has been preserved. It may also explain why since 1945 neither Soviet nor American leaders have ever been as convinced as were German leaders before 1914 of the inevitability of war.

Since the Cold War began, successive American presidents have refused to believe that the international situation was so bleak as to make war all but unavoidable. Some presidents, Harry Truman among them, expressed dire forebodings, it is true, in reaction to particularly threatening events abroad. But their pessimistic moods never lasted very long. All the Cold War presidents preferred to believe that war was avoidable and that the Soviet Union might one day mellow and become more moderate in its foreign policy goals. In the 1950s these hopes were pinned to Stalin's death and the possibility of a subsequent transformation of the Soviet political system. In the 1960s they rested upon the sobering effect that Americans expected realization of the true destructiveness of nuclear weapons to have upon Soviet leaders. Even those presidents who sustained the darkest view of Soviet intentions, Truman, Kennedy, and Reagan, still have looked forward to a time when Soviet-American relations might improve.

It is worth entertaining the hypothesis that the hope of avoiding war with the Soviet Union was a motivated bias. Because presidents who would have had to authorize the use of America's nuclear weapons did not want to believe it would ever become necessary, they revised their estimates of the probability of war downward. More intriguing still is the possibility that this bias was at least partially self-fulfilling. Prompted by moral-psychological needs and, after the development of a Soviet nuclear capability, by political-military needs as well, belief that nuclear war was unlikely may have helped maintain the peace. It made policy makers cautious rather than risk-prone and more alert than they might otherwise have been about finding ways to prevent war.

The same process may have been unfolding in the Soviet Union. At the end of World War II, Stalin believed war with capitalist adversaries, among them a revitalized Germany, inevitable in about fifteen years.[10] Stalin's death was followed by a short-lived war scare, but that soon gave way to a certain degree of optimism among his successors about the possibility of reaching an accommodation with the West.[11] Relations became strained again under Khrushchev, reaching their nadir at the time of the missile crisis. It was also Khrushchev, however, who proclaimed that war with capitalism was no longer "fatalistically inevitable."

Subsequent Soviet leaders have given every indication of believing in war as a remote rather than an immediate prospect. As statements by Andropov, Chernenko, and Gorbachev indicate, this outlook continues despite the tensions that have poisoned superpower relations since the Soviet intervention in Afghanistan and the revival of Cold War rhetoric under Ronald Reagan. Soviet leaders, too, seem to have made their belief that war could be avoided at least partially self-fulfilling.

Why, then, did the Germans, unlike the Soviets and Americans later in the century, succumb to a fatalistic acceptance of war? The answer can probably be traced to a quite different attitude toward war *qua* war. War for the Germans was certainly anxiety-provoking in the sense that it was a leap into the unknown, an event full of uncertainties and not without real human cost. Bethmann-Hollweg rightly suspected that it would sweep away more thrones than it would prop up. But the idea of war itself held few horrors for most of the German policy-making elite, many of whom conceived of war as something glorious, manly, and even spiritually uplifting. Most influential Germans assumed, moreover, that victory would have beneficial consequences for their society, by creating a mood of patriotic euphoria that would overcome the class and party differences threatening the survival of the Reich.[12]

Postwar American and Soviet leaders give no evidence they ever shared this illusion. Instead, they have viewed nuclear war as destructive of all important social and political values. Official Soviet pronouncements to the contrary were issued, to be sure, until fairly recently. They seem, however, to have been motivated by internal political considerations, and they gave no real indication of Soviet attitudes toward nuclear war. The trauma of World War II has, if anything, left Soviet citizens and leaders alike committed to avoiding any repetition of the destruction their nation suffered in that conflict.[13]

Fear provided a strong incentive for Soviet and American leaders to believe nuclear war could be avoided. Many policy makers in Washington and Moscow may even have shaded their judgments of the adversary in order to make them consonant with their desires to avoid a nuclear confrontation. By dwelling on ways in which their adversary's foreign policy might in the long run become less aggressive, and war thus unnecessary, they reduced the attraction of unduly provocative policies and at the same time magnified the perceived rewards of patience, caution, and military procrastination. Over the years success in avoiding war encouraged even greater expectations of doing so in the future.

These fundamental differences between political-military relations on the eve of World War I and those today constitute important reasons why peace has been preserved between the superpowers despite the sometimes tense state of their relations. Because of these differences, acute superpower crises are less likely today than great power crises were in the years before 1914. This is reason for guarded optimism. At the same time, however, the fact that neither superpower has ever been sufficiently desperate about its security to be willing to risk war with the other does not mean that such a state of affairs will last indefinitely. A combination of serious domestic and foreign setbacks could make either superpower much more willing to risk war. Intense strategic competition in the form of a race to deploy a space-based ballistic missile defense could also seriously aggravate tensions and mutual perceptions of threat. Leaders of one or, worse still, both superpowers could feel themselves to be in a position analogous to that of Germany or Russia in 1914. Exaggerated fears for their security could prompt one or both of them to pursue confrontatory foreign policies or even military ventures as a means of forestalling a threatening window of vulnerability.[14]

The Problem of Crisis Instability

For all the reasons this book has made clear, a grave superpower crisis, if one occurs in the future, will be very much more difficult to resolve than past crises have been. Let me emphasize again that I am talking about a confrontation of the magnitude of Cuba, not a *folie à deux* such as Angola or any of the other minor superpower crises of the last twenty years. Any clash, if managed poorly, can escalate into a more serious confrontation, of course, but the Cuban missile crisis was obviously and qualitatively different from such minor brushfires.

Perhaps the best way to summarize our findings about crisis

instability is to think of crisis management as a curve. A steep curve requires considerable energy to move an object up its slope or to transform a system from one state to another. A relatively flat curve, by contrast, permits considerable movement in response to even minimal force.

Crisis stability would be greatest in a political-military environment characterized by a steep curve, that is, one in which considerable "energy" has to be put into the system in order to move the confrontation up toward the top of the curve (in the direction of war). Such an environment would also be self-correcting; the strategic warning and response systems of the superpowers would tend to return to their day-to-day states in the aftermath of a crisis that did not lead to war.

Mutual Assured Destruction sought, but never fully achieved, crisis stability of this kind. In theory a world with two nuclear powers, each possessing a secure second-strike capability and a similar countervalue strategy, would be highly stable in a crisis. Neither side would have any incentive to preempt—quite the reverse, in fact, as preemption would be entirely suicidal. Nor would de-escalation pose a problem following the resolution of a crisis. Unfortunately neither superpower was ever reconciled to this vision of security; both sought instead to maximize their relative strategic advantage.

Several decades of superpower strategic competition have resulted in an increasingly flat crisis curve. Pressures to preempt, the possibility of loss of control, and the risk of miscalculated escalation have become more serious threats to crisis management today than they were ten or twenty years ago. All three dangers would also come into play at much lower levels of escalation now than was true in the past. Thus even a small perturbation—in the form, say, of some kind of strategic alert—could move a superpower confrontation some distance in the direction of war. Crisis stability is likely to deteriorate even further in the course of the coming decade, because of the new and more threatening strategic forces that both sides will deploy.

Structurally the political-military environment is bad enough. Politically, because superpower political leaders are insufficiently aware of the dangers, the situation may be worse still. A remarkable degree of ignorance about war plans and crisis management exists in Washington, as I have suggested. One effect of this ignorance is that many top officials give evidence they conceive of crisis management in terms of their stereotyped understanding of the Cuban missile crisis. Many give evidence of believing that crisis management consists of controllable and reversible steps up a ladder of escalation, steps taken to moderate an adversary's behavior by demonstrating resolve. Worse yet, some believe that the demonstration of resolve requires readiness to threaten the use of nuclear weapons, even to deliver on the threat "if necessary."

The difference between strategic reality and the policy maker's understanding of it constitutes a crucially important source of crisis instability. Political leaders could embark upon a course of action in a crisis—just as they did in 1914—only to learn the real implications of their policy after it was too late to do anything about it. This situation calls for a four-pronged effort to address or cope with the problem of crisis instability.

First, something must be done to reduce the pressure in a crisis to move to higher levels of strategic alert. Strategic alerts currently involve a fundamental contradiction: higher alerts may be necessary to guarantee retaliation and discourage adversarial preemption, but they also risk war by loss of control. Military officers have, for the most part, chosen to deal with this dilemma by ignoring it. Organizational compartmentalization helps them do so; few people have an overall view or responsibility for the problem. Of those who do, some seem to pretend it does not exist or can readily be surmounted. More concerned strategic analysts, military and civilian, have sought to ease the dilemma by enhancing the short-term survivability of command and control. By this means they hope to reduce the pressures that policy makers experience in a crisis to alert their strategic forces and also to minimize the risk of accidental war by making negative control feasible at higher levels of alert.[15]

Efforts of this kind are laudable but insufficient. We must go beyond the search for technical solutions and squarely confront the real source of the problem: contemporary force structures and the strategic doctrines governing their deployment and use. Until arms control measures or other arrangements reduce the ability of strategic weapons to destroy superpower leaderships in a matter of minutes, and subsequently to obliterate other vital command and control nodes, crisis instability will persist. Too narrow a focus on technical means of enhancing the survivability of C^3I could even prove detrimental, because it could divert attention from more important efforts to do something about this fundamental cause of crisis instability.

Even in the absence of arms control there is still something that can be done to help reduce crisis instability. Conventional strategic wisdom recognizes that preemption is an irrational act—unless nuclear war is inevitable. But when war becomes unavoidable, preemption becomes attractive as a way to limit the damage one suffers. Soviet strategists, and some Americans, argue that preemption would also confer a relative military advantage. My analysis indicates that preemption can *never* be a rational response to external strategic realities. Leaders can never know that war is inevitable; but they can act in ways that make their fears self-fulfilling. They did so in 1914. Military and political leaders in Russia and Germany took the fatal step of mobilizing because they exaggerated both the probability of war and the importance of getting in the first blow. For the same reasons, Soviet and American policy makers could be stampeded into making even more tragic decisions.

Purely abstract analyses of the military balance encourage leaders to believe in the "appeal" of preemption. By discounting or altogether ignoring the operational uncertainties and psychological constraints that would loom large for leaders contemplating an attack, theoretical analysis encourages officials to exaggerate the likelihood of adversarial preemption. Because analysts conceptualize war in terms of relative advantages instead of absolute costs, they also exaggerate the advantages that preemption would confer. After all, how significant would a relative advantage of one thousand nuclear weapons be in the aftermath of a nuclear exchange that destroyed hundreds of cities and hundreds of millions of people on both sides, leaving survivors exposed to high levels of radiation and perhaps to severe climatic disruption? Clearer and more sophisticated strategic thinking would expose the faulty assumptions that calculations of this sort are based on. *Political leaders, especially in the United States, must free themselves from the narrow operational criteria that currently dominate both official and public analysis of nuclear strategy.* This would in and of itself constitute an important step toward a more secure world.

The strategic significance of what goes on in the minds of political leaders brings us to the third focus of efforts to improve crisis stability. This is *the need to educate political leaders and their advisers about the details and pitfalls of war planning, crisis management, and escalation*. Ignorance of these matters, caused primarily by understandable reluctance to confront an anxiety-provoking subject, is another important structural source of crisis instability. Political ignorance was a major cause of World War I; it could help bring about World War III.

Greater presidential involvement with the details of crisis management and war planning promises a variety of positive payoffs. It could make a president more conscious of the dangers of crisis *mis*management, especially of the difficulty of maintaining effective control over alerted strategic forces. Participation in command post exercises would almost certainly expose some of the most glaring deficiencies in crisis management procedures, and perhaps they might bring home to the president just how politically irrelevant are most, if not all, of the SIOP options. Sobering encounters of this kind might even encourage more cautious foreign policies. They could also stimulate the White House to do something to make crisis and war planning more responsive to the president's political needs.

Presidents who were more strategically sophisticated could probably cope more effectively with the incredible stress that any real crisis would give rise to. Even minimal familiarity with the physical settings, individuals, and issues involved in crisis management would reduce an important element of novelty and uncertainty. It would permit the president to focus attention on the problem at hand instead of on the setting. This idea certainly does not aim to encourage the president to develop a false sense of confidence in his or her ability to manage a grave crisis, but rather it aims to reduce the stress that could seriously impair presidential judgment. It could make the difference between some kind of reasoned response and an overly emotional or ill-conceived one.

There is another possible psychological benefit. At times, it seems, American presidents have resorted to defensive avoidance in order to cope with the anxiety associated with mere contemplation of their role in a possible nuclear crisis or war. Avoidance would be much more difficult if presidents and other relevant political officials participated open-mindedly in periodic briefings and exercises. They would have to find some other way of coping with the anxiety that the experience would almost certainly engender. This need might even prove a catalyst, pushing presidents to pursue more assiduously various measures to reduce tensions in the hope of improving superpower relations and thereby reducing the likelihood of acute crisis or war. Defensive avoidance benefits only a president's subjective state; a policy that explored possible means of Soviet-American accommodation would serve the broader national interest as well.

The final and most important objective of policies aimed at war prevention must be to try to prevent acute crises altogether. Efforts must be made to manage minor crises successfully, to keep them from developing into war-threatening confrontations. The complexities of strategic operations and the continuing vulnerability of command and control impose severe limits on leaders' abilities to control them. The superpowers must learn to respect this reality and do their best to avoid confrontations with each other. American leaders must also recognize the dangers of nuclear alerts and their general inappropriateness as diplomatic signals.

In the aftermath of the Cuban missile crisis, Robert McNamara declared that there was no longer any such thing as strategy, only crisis management.[16] It is easy to understand how McNamara felt after living through such a harrowing confrontation. For crises that involve strategic escalation, however, the truth of the matter is just the reverse. Given the nature of the problems we have identified, there is unlikely to be any such thing as good nuclear crisis management. The only good strategies are those designed to prevent crises.

Governments would therefore be well-advised to devote at least as much effort to crisis prevention as they do to crisis management. Severe crisis is a late stage of the game at which to invest one's resources in war prevention. I do not recommend leaders discontinue such efforts, for they are necessary for all the reasons I have detailed. More sophisticated approaches to crisis management would also be helpful in mastering less serious confrontations. But concern for crisis management does not relieve scholars and officials from the more pressing responsibility of developing and applying the full range of political, military, and economic strategies most appropriate to crisis prevention.[17]

Reducing Crisis Instability

In chapters 2 and 3, I analyzed two major structural sources of crisis stability: the incentive to preempt, and the possibility of loss of control. Both problems have deep-rooted causes, and either could seriously impede successful resolution of any future superpower crisis involving nuclear alerts. Neither problem, I have argued, is amenable to a quick fix. Even the most ambitious efforts would succeed only partially in alleviating either source of crisis instability. The problem is made more intractable still by the fact that measures designed to cope with one source of instability can have the effect of exacerbating the other. Accordingly we need to assess all such proposals in terms of their overall effect. Identifying and evaluating these trade-offs is a formidable task, but not an impossible one.

Four kinds of uncertainties make it difficult to evaluate the likely real effect of any initiative to enhance crisis stability. The first, and least significant, of these is technical. It pertains to the uncertainties that surround the effects on connectivity of nuclear weapons, present and future antisatellite weapons, electronic measures, and sabotage. Nobody knows just how vulnerable individual components or entire communications systems are to the full range of nuclear effects. Some of these effects are still poorly understood. Electromagnetic pulse, generally recognized as the most damaging, was unknown until 1962; our knowledge of it and related effects is still incomplete. Without above-ground tests, banned for compelling political and humanitarian reasons, it is likely to remain so. The severity, range, and duration of other nuclear effects, primarily a function of the size and altitude of the detonation, are also influenced by transient atmospheric conditions. A band of uncertainty therefore will always surround damage estimates. Nevertheless, it is fair to work on the assumption that command and control is and will remain extremely vulnerable. Connectivity will not survive the opening salvos of a nuclear war.

The second component of uncertainty is psychological. In Chapter 4, I argued that it is difficult to predict how political leaders and military officials will function in a crisis. It is even more difficult to predict how they will perform in a nuclear war. Would a president or a premier respond to seemingly unambiguous warnings of attack by ordering immediate retaliation? Or would both wait, hoping to convince themselves that it was all a mistake? If they believed that an attack against their country was under way, would they act as "rational decision makers," as the SIOP supposes, and tailor their response to the nature of the attack against them? Or would they become hypervigilant, frantically switching from one course of action to another in the desperate hope of finding some way out of the crisis, perhaps seeking refuge in some more extreme defense? What about other officials down the chain of command, from the Joint Chiefs to the officers actually charged with firing the weapons? How would they react to an order that they had spent years convincing themselves would never come?

Some years ago Herman Kahn imagined the following telephone conversation between Chairman Khrushchev and a Soviet general:

> G: So you can see that if you press these three hundred buttons there is a good chance of us getting away scot-free, a small chance of us suffering moderate damage and no chance at all of us suffering as much damage as we suffered in World War II.
> K: The Americans are on a fifteen-minute alert. If they have any spies or even if we have a defector we will be destroyed!
> G: Don't worry. I have arranged to have a training count-down operation at noon every Saturday. All you have to do is pick up the telephone and give the order. You will be the only one who knows when the attack is going to take place.
> K: I don't believe it. What if some Ukrainian who is still mad at me presses one of the buttons ahead of time just to get me in trouble! . . .
> K: I still don't like it. I can imagine what will happen. I will pick up the phone and say "fire!" The officer will reply, "What did you say?" I will repeat, "Fire!" He will say, "There seems to be a bad connection. I keep hearing the word "fire." I will say, "If you don't fire I will have you boiled in oil." He will say, "I *heard you* that time. *Don't* fire! Thank you very much!"[18]

The times and the personalities have changed, but the problem stays the same. Key people in the chain of command could refuse to admit the prospect of nuclear war, as Kahn's macabre dialogue suggests. Or they could push for retaliation in the absence of compelling evidence that they were being attacked. They could even fail to react at all, overcome by the incredible stress that such a situation would generate.

Perhaps we should turn to works of fiction for insights into how surviving launch-control officers, and B-52 and submarine crews, would function in the aftermath of a nuclear attack. Would the realization that their families had probably been wiped out, their society devastated, and their own lives perhaps marked in hours or days make them bent on revenge or put them beyond caring? Novels of this genre are divided in their judgment.[19] So are the social scientists I have asked.

The constantly evolving strategic environment further complicates knowledge about the technical and organizational aspects of nuclear war. The number and characteristics of new weapons systems, the success or failure of arms control in curbing them, the extent of future efforts to make command and control more survivable, these are only a few of the developments that will significantly affect the superpowers' perception of their relative vulnerability and capability to respond to attack. Some are measurable, others certainly are not. Their collective effect, or more precisely a leader's understanding of their collective effect, is beyond our ability to predict.

Worse, a change in only one variable in this complex equation can have a significant overall effect. A decision by the United States to adopt a quick-launch doctrine, something currently being debated within the Pentagon, would enhance the country's ability to retaliate, but it would also significantly increase the chance of accidental war in a crisis. Launch on warning, as we shall see, would so change the nature of the trade-offs between the two competing objectives of crisis management that we would have to reassess all of the effects we expect from other developments or initiatives.

Finally, there is the uncertainty that surrounds Soviet doctrine and targeting. Most studies of command and control vulnerability are based on worst-case attack scenarios. They assume that any Soviet attack would attempt to disrupt U.S. connectivity to the greatest extent possible. And indeed, this is probably a more accurate portrayal of strategic reality than the worst-case scenarios regularly used to assess force vulnerabilities, because we know the Soviets place great emphasis in doctrine and practice on destroying the command and control nodes of their adversary. Nevertheless, all worst-case analyses make unrealistic assumptions about the skill of the other side's military planners and the operational reliability of the opponent's command structure and forces. It is entirely possible that a Soviet attack, even one specifically designed to destroy American command and control, would fail to achieve its objective.

The Soviets could also change their targeting strategy. In the 1960s, when the Soviet Union's strategic forces were distinctly inferior in quantity and quality, it made sense for Moscow to target the American command and control system. This strategy must have been all the more tempting at the time because of the remarkable American lack of concern for U.S. vulnerabilities. But current

U.S. efforts to enhance connectivity may well lead Soviet military authorities to conclude that even the best-coordinated attack, even one that destroys American command and control, has little prospect of preventing massive retaliation. To the extent the Soviets think so, they would have a strong incentive to reconsider their own approach to nuclear war. They might just conclude that limited strike options, which Soviet doctrine currently rejects as unrealistic, offer a better chance of warding off substantial damage to their homeland.

It is certainly difficult to define and measure what affects trade-offs between policies designed to ensure retaliation and those aimed at preventing loss of control. But such difficulties should not discourage efforts toward this end. Rather, they point to the need to reduce as far as possible the levels of uncertainty that currently surround such matters. Definitive discussion of the trade-offs between the two sometimes contradictory goals of crisis management must await the findings of further research. Our exploration of the subject nevertheless encourages and permits some recommendations.

The Need for More Survivable C³

Anything that would enhance the short-term survivability of strategic command and control would also enhance crisis stability. "Short-term" in this context means the first hour of a nuclear war, the time in which a retaliatory strike would be launched. More durable command and control would go a long way toward reducing pressures on policy makers to preempt. It might even permit the retention of negative control well into an acute crisis or the opening phase of a war.[20] Many measures could be put into effect without in any way raising the risks of loss of control; some would actually improve control.[21] A case in point is the hardening and proliferation of communication channels between the National Command Authority and the unified and specified commands.

Efforts to improve the survivability of command and control are already under way; some of them I described in Chapter 2. Progress toward this goal has nevertheless been slow. Command and control improvements cut across service boundaries, and all of the services have given them low priority. They see command and control as peripheral to their central missions and interests. Better command and control also raises the prospect of greater integration of the services and tighter civilian control over them, constituting another source of resistance.[22]

Traditional service neglect of command and control vulnerabilities defies simple bureaucratic explanation. Connectivity is essential to SAC's primary mission. If communications links between the NCA and the strategic forces do not function, SAC's missiles and bombers cannot receive the authorization they need to proceed to their targets. SAC's evident lack of concern with this problem for so many years could be interpreted as evidence of its commitment, firmly entrenched since the days of Curtis LeMay, to use its forces preemptively.[23] Connectivity is irrelevant to the kind of attack LeMay had in mind—a massive first strike using all available forces.

An all-out attack has not been national policy since SIOP-63 came into effect on 1 July 1962. Nor is preemption; the United States is publicly committed to riding out a first strike before retaliating. SAC officials nevertheless have from time to time given indications that this is not what they expect to do. If SAC is committed to launch on warning or launch under attack, it has a strong incentive to oppose command and control improvements that would make riding out an attack more feasible.

There is another hypothesis to consider. The reluctance that SAC, navy, and Defense Department officials have shown over the years to delve too deeply into the vulnerability of command and control may have been a function of their doubts about their ability to do very much about it. Instead, they practiced collective denial. Institutional and psychological motivations to ignore command and control vulnerability could have reinforced each other. If so, these biases are important barriers that must be overcome by policy makers concerned with crisis stability. The persistence of the problem was illustrated by the experience of the Reagan administration, which committed $18 billion to improve the survivability of strategic command and control. The services, nevertheless, resisted increasing their actual expenditure on C^3.

The Dangers of Quick Launch

In contrast to the Soviet Union, the United States is publicly committed to riding out a first strike; Washington will not retaliate until after a substantial number of Soviet weapons have actually exploded on or over U.S. territory. This doctrine was adopted in the early 1960s, when it appeared a judicious means of preventing accidental war. "Launch after impact" guaranteed that radar operators would not fire unrecallable missiles in response to their mistaking migrating geese for Soviet missiles.[24]

Some clarification of the point is in order. Although the United States is publicly committed to ride out a first strike, it is *not* committed to refrain from the first use of nuclear weapons. The historical development of U.S. nuclear force structure and doctrine and one of its longstanding rationales, "extended deterrence," are predicated upon the possibility of first use of nuclear weapons. If retaliation against a large-scale Soviet nuclear attack on the United States were the sole raison d'être of American nuclear weapons, then the first fifteen years of U.S. nuclear deployments would be inexplicable. Not until sometime in the 1960s was the Soviet Union able to mount a large-scale attack on the United States. Before that time American nuclear weapons would have been used in response to lesser provocations; in Europe and maybe the Persian Gulf, Korea, and other places, they still would be. This is the basis of "extended deterrence" and the stated justification of the ongoing deployment of a new generation of American nuclear weapons in Europe. The United States continues to reserve the right to use nuclear weapons first. It has never issued any statement to the effect that it would avoid attacking Soviet territory with these weapons until or unless the USSR had first carried out a nuclear attack against the United States.

In recent years, some analysts have urged the United States to adopt a posture of launch under attack or even launch on warning.[25] The two concepts are often used interchangeably, but it is useful to distinguish between them. LOW refers to an American nuclear strike launched after U.S. sensors have detected incoming Soviet missiles, LUA to U.S. retaliation after the first Soviet warheads have detonated on or over American territory.

The goal of launch on warning is to allow the United States to launch its missiles and bombers before any Soviet warheads strike the United States and disrupt connectivity. As a strategy LOW is extremely demanding, perhaps even impossible, because only minutes—possibly as few as seven—would elapse between the detection of SLBM launches and the arrival of warheads over

Washington, D.C., and east coast naval and air bases. In all probability, only some missiles or bombers could be launched or sent aloft before Soviet warheads began to detonate.[26] LOW therefore shades into LUA, which in turn spills over into retaliation after impact. All three strategies represent distinct but overlapping segments of a continuum. Any choice among them involves a tradeoff between the somewhat contradictory goals of guaranteeing retaliation and avoiding accidental war.

Despite its official commitment to ride out a first strike, the United States is reported to have had some quick-launch capability ever since the creation in 1960 of the first Single Integrated Operational Plan. The plan apparently incorporated LOW and LUA options.[27] For a few years in the late 1970s the LOW option received more emphasis, although high-ranking military officers remained skeptical of its feasibility for both technical and psychological reasons.[28] More recently the Reagan administration has rekindled interest in LOW, seeing in it a possible solution to the problem of Minuteman vulnerability. In April 1983, testifying before a congressional committee on MX deployment, Secretary of Defense Caspar Weinberger and Chairman of the Joint Chiefs of Staff John W. Vessey, Jr., indicated that the United States might not be willing to absorb a Soviet attack before retaliating. The two officials refused to elaborate on their remarks in public, but their import was clear. The implications, moreover, were consistent with rumors circulating at the time that the administration was planning to accompany MX deployment with a LUA or LOW posture.[29]

Quick-launch postures are, to some observers, a substitute for preemption. They do away with the pressure to preempt to the extent that they make policy makers more confident about their ability to retaliate. At first glance, they also appear to have another virtue, that of finessing the principal drawback to preemption: the possibility, even the likelihood, of making one's fear of war self-fulfilling. With a policy of launch on warning or under attack, no decision need be made to launch until *after* sensors report that the adversary has actually begun an attack.

LOW and LUA can also undercut whatever incentive an adversary has to preempt. Preemption is attractive only if it holds out the prospect of meaningful victory or significant damage limitation. It loses its appeal if the other side will still be able to retaliate effectively. Preemption against an adversary with a quick-launch capability, however, would only succeed in making nuclear war, and with it one's own destruction, a certainty.

Quick-launch postures are superficially appealing for these two reasons. But they also have serious drawbacks. For a start, they entail a significantly higher risk of accidental war. Bombers and command aircraft can be sent aloft and later recalled, but ICBMs cannot be recalled, nor, at the present time, can they be disarmed after being launched.[30] LUA, and LOW even more, therefore demand an organizational capacity to make an almost instantaneous decision to respond to attack coupled with an entirely foolproof mechanism to safeguard against false alerts. Both requirements are unrealistic. There is no way of guaranteeing that a complex system of sensors, computers, related software, and human operators can function error-free all of the time. In Chapter 3, I discussed some of the kinds of errors that could occur in the context of today's strategic doctrine and command and control architecture. A hair trigger—and this is a fair description of quick-launch options—would multiply the probability of error by several orders of magnitude.

The danger of accidents aside, some knowledgeable critics have expressed serious doubts as to whether quick-launch procedures could be made to work. ICBMs would have to be launched within five minutes of the time a Soviet attack was detected in order to prevent their destruction from blast or atmospheric effects. LUA, which imposes less severe requirements, would allow up to twenty minutes for launch, but it would safeguard ICBMs only against destruction in their silos. To avoid pin-down, the missiles would have to be fired within twelve minutes of detection. Launch itself is, in fact, only the last and by no means the most difficult operation associated with a quick-launch posture. Many things have to happen before ICBMs are launched. Tactical warning of attack has to be received, verified, and sent up the chain of command; the president has to be alerted and convinced that an attack is under way, and the presidential order to retaliate has to reach the secretary of defense and be passed to the chairman of the Joint Chiefs. The go-code then has to be transmitted to the missile crews, who would first confirm it and then, if they believed it, turn their keys.[31]

All of these procedures depend upon a smoothly functioning command and control system. But in all likelihood, critical components of the system would come under attack before the ICBM silos would. Any slowing or disruption of decision making and communication, whether by organizational malfunction or by nuclear assault, would impede the practical functioning of any quick-launch option. It is with these problems in mind that John Steinbruner dismisses both LOW and LUA as entirely unrealistic options:

> With only partially discriminated warning signals and disrupted communications, normal command procedures encumbered by negative-control requirements would almost certainly delay the protective firing of U.S. missiles, and improvised attempts to expedite the process would risk perverse success: slightly mistimed launches that would expose U.S. missiles to destruction in their boost phase. Given current military capabilities, the prevailing judgment that launch under attack is impractical to the point of impossibility seems to be warranted.[32]

The time constraints associated with quick-launch procedures also constitute a serious political problem. They require that the decision to retaliate be made nearly instantaneously with the receipt of warning of attack. The president would be unable to consult at length, if at all, with trusted advisers. The commander-in-chief would be the prisoner of the military or, more accurately, of a complex alert system of which he probably had little understanding and which he would not trust if he did. Launch on warning would, in effect, delegate the decision to go to war to some computer algorithm. It seems inconceivable that any incumbent, even one convinced of the technical feasibility of launch on warning or launch under attack, would voluntarily renounce authority in this way over the presidency's most awesome constitutional power.[33]

There is another drawback to consider. Reliance on quick-launch procedures creates a strong incentive in any serious crisis to move fairly quickly up the alert ladder. Speed is essential to bring strategic forces and their command and control up to the state of readiness required to execute a LOW or LUA option. Some observers might object that American forces are already kept at a high state of day-to-day readiness; 90 percent of the ICBM force, 30 percent of the bombers, and about 50 percent of the submarines are normally on alert.[34] These forces, it is true, would be sufficient to go to war with. But almost all SIOP options are better served by forces at full wartime readiness.[35]

Retaliation also depends on fully functioning command and control. Like the strategic forces, the command and control structure is kept at a reasonably high level of day-to-day readiness. However, full wartime readiness would bring command and control networks up to an even higher state of readiness, which would presumably be important for quick-launch options. It would also alert all personnel to the gravity of the situation, something that policy makers would want to do if they considered an attack against their country even remotely likely.

Force generation by one superpower, even if carried out for avowedly defensive purposes, would arouse fears of preemption in the capital of the other. Its leaders would likely generate their forces in response, if only to discourage preemption. But their action could make preemption more likely, by convincing the superpower that had gone on alert first that war was now all but unavoidable. Even if matters did not go that far, mutual force generation would seriously aggravate a crisis, making it that much more difficult to resolve. It also carries with it, as Chapter 3 demonstrated, grave risks of loss of control.

Quick-launch options, for all of these reasons, do not offer a technically viable or a politically feasible solution to the problem of strategic vulnerability. They can nevertheless make some contribution to crisis stability. If each superpower believes that the other has a quick-launch option it is prepared to use, then it will be reluctant to consider preemption.[36] If, at the same time, neither superpower actually relies on its quick-launch option, both can avoid the danger of accidental war associated with launch under attack, and even more with launch on warning. But they can do so, of course, only if they refrain from force generation in a crisis. Crisis stability therefore requires a launch-under-attack option that would use only those forces on day-to-day alert.

The present strategic environment approximates, to some degree, the situation described above. Each superpower gives evidence of believing in the other's capability to exercise some kind of quick-launch option. If this is so, any move toward greater reliance on quick-launch options can only diminish national security.

Upgrading Early Warning

Crisis instability could be somewhat reduced by upgrading the quality and reliability of the early warning systems of both superpowers. As of 1987 the Soviets still lack effective eyes and ears to provide them with timely and accurate information about an

American attack. Their efforts to provide constant real-time surveillance of U.S. missile silos have encountered repeated technical difficulties. To make matters worse, their over-the-horizon radars do not cover all the compass points from which submarine- and air-launched delivery systems could attack them. Nor are these surveillance systems linked to information-processing systems capable of distinguishing reliably between a limited and an all-out attack or between a counterforce and a countercity attack.

American systems are considerably more sophisticated, and they have some capability to predict the damage likely to result from an incoming attack. It is rumored, however, that they too can be overwhelmed by an attack of more than several dozen warheads. Such would almost certainly be the case in any Soviet strike.

Inadequate warning systems are dangerous in two ways. False or misleading information can prompt either side to launch a strike in the mistaken belief that it is retaliating. We have already examined a scenario of this kind involving the United States. In point of fact the Soviet Union is much more likely to commit this kind of error, both because its warning system is less reliable and because it has a greater apparent reliance on quick-launch options.

If war did break out, inadequate warning systems would tend to push both parties toward total engagement. A side incapable of distinguishing the size and nature of an incoming attack is almost certain to assume the worst. It will respond accordingly. If nuclear "firebreaks" are feasible at all, they depend upon the attacked nation gaining early knowledge about which thresholds the attackers have crossed and which they have chosen to respect. This discrimination requires accurate data about the number and trajectories of incoming warheads, which in turn depend upon reliable and sophisticated radar and satellite surveillance and state-of-the-art information processing. Even so, the short span between the separation of warheads from missiles and actual impact—about eight minutes at most—does not leave much time for data processing, political reflection, decision, and implementation.

The danger of accidental war in a crisis seems partially amenable to redress by the development of more reliable and thorough warning systems. It would be in the interests of both superpowers to cooperate toward this end. One possibility is the creation of an international monitoring agency to provide real-time surveillance of both sides' missile fields.[37] Located in a neutral country, say

Sweden, such an agency could be provided by the superpowers, or perhaps by the Japanese, with the hardware (satellites, radars, computer systems) and software needed to carry out this task. It might include representatives of both superpowers on its staff. The agency would need secure "hot lines" to Washington and Moscow, to report on any changes in the state of readiness of either superpower's strategic forces. It might, if Moscow and Washington agreed, also monitor theater nuclear systems.

An international monitoring agency would supplement, not replace, the surveillance that both superpowers already carry out. It would have no role in managing actual crises, only in providing reliable information to policy makers on both sides about the alert levels of their respective forces.[38] The benefits would be both strategic and political. In a crisis, detailed and accurate information can serve as an effective antidote to the baseless fears that confrontations of this kind so often spawn. The proposed agency would also act as an important check on both superpowers' early warning systems, minimizing the chances of a tragic mistake stemming from hardware failure or malfunction. Finally, an independent source of information on alert levels would make it more difficult for military leaders on either side to exaggerate willfully the warlike preparations of their adversary in order to stampede their own leaders into higher levels of alert.

Among its important political benefits, the creation of an international monitoring agency would be a notable confidence-building measure, because it would indicate some degree of mutual willingness to renounce first-strike options. Thus it might help moderate the more extreme judgments each side holds about the other's intentions. Such a shift in threat assessment would have implications far beyond the narrow, if important, arena of crisis management.

The Need for Arms Control

We cannot look at command and control problems in isolation from force structures and doctrines for their use. Weapons that put the command and control of either side at risk are destabilizing.[39] SSBNs on offshore patrol, Pershing IIs, and antisatellite weapons are cases in point. By directly threatening important command centers, they compress warning and response time, thereby intensifying the incentive to rely on LOW, LUA, or preemption. Weapons of this kind, some already deployed by the superpowers and others scheduled to come on line in the near future, negate much of the value of measures now under way to improve connectivity. This disturbing reality constitutes a compelling reason for arms control.

Arms control could go a long way toward facilitating connectivity and thereby easing crisis instability. One important step would be the creation of "keep-out zones" around both superpowers' capitals, in order to protect their leaderships from sudden destruction.[40] A suitable radius for a keep-out zone might be 2500 kilometers. It would compel the United States to withdraw its P-II missiles from West Germany; in return, the Soviet Union would have to keep its SSBNs and cruise-missile-firing submarines well away from the east coast of the United States. Other proposals to consider include consultations on alert procedures, a prohibition on testing SSBNs in depressed trajectories, a ban on forward deployment of cruise missiles, the creation of undersea command posts and sanctuaries for SSBNs, the creation of risk reduction centers, and the improvement of the hot line between the superpower capitals. Most of these suggestions have already been voiced, and they have been subjected to a fair degree of analysis.[41]

Some of these measures one could construe as favoring one superpower over the other. Washington is more vulnerable than Moscow, for example, to destruction by missiles fired in depressed trajectories from offshore submarines. Soviet command and control nodes, on the other hand, are more exposed to destruction by theater nuclear forces, as many of them are within range of P-IIs based in West Germany. However, any attempt to ease Moscow's dilemma by withdrawing the P-IIs would create a crisis of confidence among European conservatives unless it was made part of a broader agreement from which they too stood to gain.

Political obstacles aside—and these are certainly substantial—an agreement of this kind would be difficult to reach because it transcends the traditional compartmentalization of theater and strategic weapons systems. In point of fact, these weapons ought to be treated together in arms control talks, because they are linked in all kinds of ways (and not just with regard to command and control). Alternatively, one could give command and control issues their own negotiating forum. The main advantage would be to impart greater salience to the problem of connectivity. Negotiations about command and control could also take up doctrinal and other questions that affect crisis stability but are currently excluded from arms control talks.[42]

Although concern is growing in both the United States and the Soviet Union about the character and development of each other's war plans, no effort has been made to make those plans the subject of negotiations. The SALT process and subsequent arms control talks have restricted themselves to numbers and kinds of weapons deployed or under development and the procedures for verifying any agreement. At the time the American side saw the narrow engineering mentality that shaped the SALT negotiations as essential to success. In retrospect it may also have been its greatest failing. If and when serious arms talks resume, war plans should be featured on the agenda.

The first step in this direction would be for each superpower to formulate the kinds of verifiable deployments and command and control procedures that the other would have to initiate in preparation for a disarming strike. Of special interest would be those steps not associated, individually or collectively, with other possible kinds of attack. As much of this information, if it is known at all, is classified, this question is difficult to discuss in detail in the open literature. There are, however, distinctive preparations associated with a disarming strike which are on the whole identifiable and probably subject to monitoring. One obvious example pertains to the capability of firing large numbers of missiles in salvos, currently somewhat limited on both sides by the safeguards that each has put in place to prevent accidental or unauthorized launches. Another is the kinds of weapons systems that are deployed.

It has been alleged that SALT II was possible because it banned nothing that either superpower felt strongly about. Instead, it legitimized weapons programs they were interested in or to which they were already committed. An agreement that regulated attack scenarios and associated deployments would probably have to be constructed along similar lines. It could aspire to ban or regulate scenarios that neither superpower found particularly attractive. Such an agreement might be possible for an ironic reason: each superpower seems to fear most what its adversary appears least likely to do.

American defense planning takes as its baseline the need to

prevent a nuclear Pearl Harbor, which so many American strategists have for decades seen as the most likely way for World War III to begin. Thus the United States has built a highly redundant strategic arsenal, divides its forces among land-based missiles, submarines, and bombers, and maintains all of its forces at relatively high states of readiness. All of these measures are designed to guarantee a potent second-strike capability in the aftermath of a premeditated attack.

Soviet defense planners for their part seem to fear a repetition of their World War II experience—a massive American assault that would utterly devastate their country. Through the early 1960s U.S. war plans aimed to achieve precisely this goal; their objective was to wipe out sufficient population and industry to destroy the Soviet Union as a functioning country. Since then, however, American war planners have increasingly moved toward more selective and limited options, some of them designed, as far as feasible, to leave the Soviet population and industrial base intact.[43] It is in keeping with this strategy that the overall yield of the American arsenal has declined, even though the number of warheads in it has greatly increased.

This change was brought about partly by technological developments that permit the targeting of discrete and even hardened military assets. But it was also a response to the Soviet development of a potent retaliatory capability. The last thing American planners wanted to do was to launch an all-out attack against the Soviet Union which would provoke a similar assault against the United States. The U.S. objective became one of finding ways of keeping nuclear war limited, by building firebreaks between the first use of nuclear weapons on either side and all-out nuclear war.

An agreement that bans some of the routines and deployments associated with the kinds of attack which each superpower fears most would not only reduce crisis instability but also significantly lessen superpower tensions. Its feasibility, like that of any arms control agreement, hinges on the development of adequate means for its verification. Verification in this instance would unquestionably require some kind of cooperation; perhaps both sides would have to agree to transmit *en clair* all internal communications associated with exercises or other related organizational activities. Such an agreement would be analogous to ceasing to encrypt some missile test data, one of the stipulations of the SALT agreement.

Redesigning Strategic Alerts

An important if less visionary measure for reducing crisis instability involves the reorganization of the alerting procedures of the superpowers. The objective would be to minimize the chances of loss of control and, if possible, to make defensive measures appear less threatening. The United States has five levels of strategic alert, each associated with a different level of readiness for both conventional and strategic forces. What little is known about alerting procedures suggests they may not be at all suited to the needs of crisis management (hardly surprising, because they were not designed with this end in mind). Authorization of a higher DefCon prompts greater readiness across the board, throughout the strategic and conventional commands. Procedures exist to put specific elements of the strategic or conventional forces at levels of alert even higher than the authorized DefCon; military commanders, acting on their own authority, can even use them. But there does not seem to be any mechanism for keeping forces at a level of readiness *lower* than the authorized DefCon.

It would make sense to isolate certain kinds of forces and missions and to devise procedures for regulating their readiness individually. "Withholds" could be built into alerting options just as they are now incorporated into the SIOP with respect to certain kinds of targets. A move to a higher DefCon then would not of necessity move all forces, or even all those forces within a certain category, to a higher state of readiness.[44] Political leaders must be made aware of these possibilities. In a crisis they could exploit the resulting flexibility to reduce the likelihood of preemption or loss of control.

The most important organizational step in this connection would be to redesign the alerting system. What is needed is a system with the capability of bringing command and control up to a high state of readiness without automatically doing the same for strategic forces. Such a move would significantly enhance the likelihood of retaliation in the aftermath of an attack without at the same time conveying to an adversary the same sense of threat as a full-scale alert would. Greater flexibility with respect to readiness measures would also permit a response more carefully tailored to the nature and locale of the threat. This reform would minimize the risk of loss of control by permitting more discrete and carefully managed military preparations.

Preserve the ABM Treaty

Crisis stability requires the maintenance, or better yet the strengthening, of the Anti-Ballistic Missile Treaty. Ballistic missile defense (BMD), the goal of President Reagan's Strategic Defense Initiative, would intensify mutual pressures to preempt in a crisis. BMD systems would be even more vulnerable than land-based missiles and command and control. All of the proposed BMD weapons would be designed to operate in space or would require some space-based components. These would include sensors to detect enemy ICBM launches and, in the case of lasers, large numbers of mirrors to reflect energy beams generated on earth. Placed in low orbits, these components would be vulnerable to attack by ground-based lasers or missiles launched from fighter planes.

BMD systems or components in higher, geosynchronous orbits could be destroyed by collision or by nuclear explosions set off in space to destroy them or degrade the sensors they need to track ICBMs. Either superpower could also orbit "space mines," detonating them whenever desired by remote control. If both sides have BMDs, either system could be used to kill the other. Even minimally effective BMDs could destroy the adversary's communication, surveillance, and other satellites with great rapidity.

This vulnerability of BMDs means that the principal utility of this kind of weapon is as an important component of a first-strike strategy. As a truly defensive weapon, a BMD would have to destroy all or almost all attacking missiles. This, for a variety of technical reasons, is simply not feasible.[45] A partially effective BMD, however, might be more effective against the second strike of an enemy many or most of whose weapons had already been destroyed. A BMD therefore not only increases the pressure on political leaders to launch a first strike in a serious crisis but also makes such an attack a much more attractive prospect. This reality, known to both sides, would generate even greater pressures to launch a first strike for fear that the adversary was about to do so.

There is a close link, technologically and strategically, between space-based BMD systems and antisatellite weapons. ASATs provide the most obvious countermeasure to space-based BMD systems or components. The fact that ASAT and BMD are based in part on the same technology means that the success or failure of efforts to ban ASATs has major implications for the control of BMD development. Unrestricted ASAT development would undermine the prohibition in the ABM Treaty on space-based missile defenses, while restrictions or a ban on ASATs would reinforce the treaty and greatly impede progress toward space-based defenses. An ASAT ban would also avoid the hazards to crisis stability which advanced ASATs are likely to pose.

Conclusion

The measures I have described, together with a more skeptical attitude toward the putative advantages of preemption, would go some way toward reducing crisis instability. However, retaliation and control will remain somewhat contradictory goals; the conflict between them cannot be completely reconciled or finessed. Beyond a certain point, measures designed to guarantee retaliation will increase the risk of loss of control, and vice versa. This inescapable structural reality makes an acute crisis very dangerous, something that superpower leaders must do their best to avoid. Failing that, they must attempt to manage any crisis at the lowest levels of escalation possible.

The Next Strategic Frontier

The focus of American strategic concern has undergone two important transformations in the course of the Cold War. The catalyst for the first of these was Albert Wohlstetter's now famous Rand study of bomber basing. By making apparent just how vulnerable the European-based bombers were to a Soviet first strike, Wohlstetter sensitized the strategic community to the need for enough strategic forces to survive in order to ensure effective retaliation.

This concern—some would call it a fixation—has dominated American strategic thought and policy down to the present day. In the early 1960s it provided the incentive for deploying an increasing percentage of America's strategic forces at sea. A decade later it found expression in doubts about the survivability of land-based missiles, allegedly vulnerable to the newly MIRVed SS-9s. Concern became acute in the late 1970s, in response to Soviet deployment of the larger and more accurate SS-18. It spawned frightened talk of a "window of vulnerability" and of the urgent need for a more survivable land-based ICBM. The MX missile, envisaged as the solution to this problem, will be the most expensive and destabilizing weapon of the decade—if it is ever deployed. It has also been the most controversial.

For almost three decades, this concern about force vulnerability obscured the even greater vulnerability of U.S. command and control. The strategic community has belatedly become aware of the extent of this vulnerability, thanks to the pathbreaking studies of Desmond Ball, John Steinbruner, Paul Bracken, and Bruce Blair. Blair asserts that the Soviets, by contrast, have long been aware that command and control was America's strategic Achilles heel; Moscow made it a primary target for its nuclear forces as long ago as the 1960s. U.S. retaliatory capability was actually far from certain, according to Blair, throughout the years when successive administrations were convinced of their country's strategic superiority.[46]

The collective effort of these scholars has done much to make the strategic community more aware of the importance of survivable command and control. Command and control issues have attained increasing prominence in the strategic literature, although they have not engaged the attention of policy makers to nearly the same degree. The Carter and Reagan administrations nevertheless displayed more concern for the problem than their predecessors did. It is reasonable to suppose that greater efforts will be made when the implications of command and control vulnerability become more widely known to the Congress. We are thus in the midst of a second shift in strategic emphasis; sooner or later there is likely to be as much concern for command and control vulnerability as there is for force vulnerability.

This is, on the whole, a salutary development. In the past policy makers and armchair strategists generally exaggerated force vulnerabilities while ignoring those of command and control. Some kind of corrective has long been overdue. We must be careful, however, not to go too far in the other direction. Concern for survivable command and control should not blind us to the dangers of loss of control. Policy makers must eschew quick fixes or relatively inexpensive measures that enhance the prospects of retaliation at the expense of maintaining control over strategic forces in a crisis. This will mean more expensive C³I programs. But in terms of its payoff for crisis stability, the money will be well worth it.

Command and control is by no means the final strategic frontier. The history of twentieth-century crises tell us that human performance is the single most important component of crisis management.[47] Human limitations may also constitute the most important source of strategic vulnerability.

Retaliation and restraint both depend on responsible and timely behavior at every level of the chain of command, from the president down to missile launch-control officers and bomber and submarine crews. Our knowledge of the psychological processes that influence human behavior indicates that this is a remarkably optimistic assumption. Such a judgment is not based on inference alone; past crises reveal that severe stress is a significant impediment to good decision making; it may induce erratic or irrational behavior or even paralysis. The stress generated by a nuclear crisis is almost certain to surpass that of previous crises, both because of the unprecedented destructiveness of nuclear weapons and because of the kind of time pressures political leaders will face. It would be surprising if these conditions did not result in unexpected and quite dysfunctional behavior throughout the chain of command.

So far, this problem has barely been recognized in the literature. Analysts display sensitivity toward the technical shortcomings of strategic forces and their command and control but almost none regarding the limitations of the people who are ultimately responsible for making both of them work. Ignorance—perhaps avoidance—of this problem is in every way reminiscent of the earlier lack of concern for command and control vulnerabilities. Yet the human factor constitutes an important strategic vulnerability, and it is a possible cause of loss of control in crisis. Indeed, the human factor is the most significant strategic frontier remaining to be explored. We need to know much more about the limits and potentialities of people subjected to complex problems and acute stress. That greater understanding will prove as important as insights from other areas of research to our efforts to augment crisis stability.

Notes

1. *A Dangerous Illusion*

1. Theodore C. Sorensen, *Kennedy* (New York: Harper & Row, 1965), p. 705.

2. Personal communications with Robert McNamara and McGeorge Bundy.

3. Dean Acheson, "Homage to Plain Dumb Luck," *Esquire*, February 1969, pp. 44–46, 76–77; Edward Weintal and Charles Bartlett, *Facing the Brink: An Intimate Study of Crisis Diplomacy* (New York: Scribner's, 1967), pp. 54–55, quoting the official crisis postmortem prepared by Walt Rostow and Paul Nitze in February 1963.

4. See Henry Kissinger, *Years of Upheaval* (Boston: Little, Brown, 1982), pp. 575–90; Raymond L. Garthoff, *Détente and Confrontation: American-Soviet Relations from Nixon to Reagan* (Washington, D.C.: Brookings, 1985), pp. 374–84; and Scott Sagan, "Nuclear Alerts and Crisis Management," *International Security* 9 (Spring 1985), pp. 99–139.

5. See, for example, Thomas Schelling, *Arms and Influence* (New Haven: Yale University Press, 1966); Albert and Roberta Wohlstetter, *Controlling the Risks in Cuba*, Adelphi Paper no. 17 (London: International Institute of Strategic Studies, 1965); and Irving L. Janis, *Victims of Groupthink* (Boston: Houghton Mifflin, 1972).

6. U.S. alerting procedure and U.S. and Soviet military preparations during the Cuban missile crisis are discussed in detail in Chapter 2.

7. See Richard Ned Lebow, *Between Peace and War: The Nature of International Crisis* (Baltimore: Johns Hopkins University Press, 1981). For other characterizations of good decision-making environments, see Irving L. Janis, *Groupthink*, rev. ed. (Boston: Houghton Mifflin, 1982), pp. 260–71, and Alexander L. George, *Presidential Decisionmaking in Foreign Policy: The Effective Use of Information and Advice* (Boulder, Colo.: Westview, 1980), p. 10.

8. John Steinbruner, "Nuclear Decapitation," *Foreign Policy* no. 45 (Winter 1981–82), pp. 16–28; Desmond Ball, *Can Nuclear War Be Controlled?* Adelphi Paper no. 169 (London: International Institute of Strategic Studies, 1981); Paul Bracken, *The Command and Control of Nuclear Forces* (New Haven: Yale University Press, 1983); Bruce G. Blair, *Strategic Command and Control: Redefining the Nuclear Threat* (Washington, D.C.: Brookings, 1985).

9. Even Richard Pipes, strategic Cassandra *par excellence*, does not believe that World War III would begin this way. He told Michael Getler of the *Washington Post:*

"I personally do not believe that the Soviet Union would launch a preemptive [*sic*] nuclear strike against the United States out of the blue, no matter what the balance of power is. I just don't think that is in the cards." 22 July 1982, p. 8.

10. Douglas Terman, *First Strike* (New York: Pocket, 1980), is typical of this genre. The Soviets continually feed all information pertaining to the relative military strength of the superpowers into a computer that determines the numerical correlation of forces between them. The evil masters of the Kremlin are committed to launching a first strike when they achieve a given level of advantage.

11. H. R. Haldeman, *The Ends of Power* (New York: Times, 1978), p. 90, reports that during the SALT negotiations the Soviets expressed concern that China would be tempted to try to provoke a nuclear war between the superpowers.

12. Early theoretical treatments of preemption include Thomas C. Schelling, *The Strategy of Conflict* (Cambridge: Harvard University Press, 1960), pp. 207–54, and *Arms and Influence* (New Haven: Yale University Press, 1966), pp. 221–48; Glenn H. Snyder, *Deterrence and Defense* (Princeton: Princeton University Press, 1961), pp. 97–114. The quotation is from *Strategy of Conflict*, p. 207. Schelling's view of this problem has not changed. In a recent article, "Confidence in Crisis," *International Security* 8 (Spring 1984), pp. 55–66, he asserts that "the greatest danger in a nuclear crisis is the potential self-fulfilling prophecy that mutual suspicion has reached an intolerable level at which preemptive action is inevitable."

13. Lawrence Freedman, *The Evolution of Nuclear Strategy* (New York: St. Martin's, 1981), pp. 163–65, the most highly respected work of its kind, limits its discussion of crisis stability entirely to an analysis of the pressures for preemption.

14. Luigi Albertini, *The Origins of the War of 1914*, 3 vols., trans. Isabella M. Massey (London: Oxford University Press, 1952), 2: 253; A. J. P. Taylor, *The Struggle for the Mastery of Europe, 1848–1918* (New York: Oxford University Press, 1969), p. 444. The most extreme claim for the role of loss of control in bringing about World War I is F. H. Hinsley, *Power and the Pursuit of Peace* (Cambridge: Cambridge University Press, 1967), p. 296, who argues that "the dice had been set rolling for all the Powers well before Russia mobilized. . . . All the evidence goes to show that the beginning of the crisis, which has been studied so largely with a view to discovering and distributing human responsibility, was one of those moments in history when events passed beyond men's control."

15. Quoted in W. J. Broad, "Experts Say Satellites Can Detect Soviet War Steps," *New York Times*, 25 January 1985, p. A12. For a fuller treatment of the subject see Bracken's pathbreaking *Command and Control of Nuclear Forces.*

16. In the case of Korea, scholarly opinion is nearly unanimous. See, for example, Allen S. Whiting, *China Crosses the Yalu: The Decision to Enter the Korean War* (New York: Macmillan, 1960); Martin Lichterman, "To the Yalu and Back," in Harold Stein, ed., *American Civil-Military Decisions* (New York: Twentieth Century Fund, 1963), pp. 569–642; Alexander L. George and Richard Smoke, *Deterrence in American Foreign Policy: Theory and Practice* (New York: Columbia University Press, 1974), pp. 184–234; Lebow, *Between Peace and War*, pp. 148–228. With regard to 1967, opinion is divided. Some analysts of the crisis believe that Nasser was intent on war from the outset. See, for example, Theodore Draper, *Israel and World Politics: Roots of the Third Arab-Israeli War* (New York: Viking, 1967), pp. 70–82, who argues that Nasser deliberately set up a situation that he knew would compel Israel to attack. More prevalent is the view that Nasser was not out to provoke a war but misjudged Israel's response; he may also have lost control over events to domestic opinion after he expelled the United Nations Emergency Force from the Sinai. So argues Nadav

Safran, *From War to War: The Arab-Israeli Confrontation, 1948–1967* (New York: Pegasus, 1969), pp. 271–302, and Walter Laqueur, *The Road to War: The Origin and Aftermath of the Arab-Israeli Conflict of 1967–68* (Harmondsworth: Penguin, 1969), pp. 122–25, 254–72. On the Falklands war, see Richard Ned Lebow, "Miscalculation in the South Atlantic: The Origins of the Falklands War," in Robert Jervis, Lebow, and Janice Gross Stein, *Psychology and Deterrence* (Baltimore: Johns Hopkins University Press, 1985), pp. 89–124.

2. *Preemption*

1. Lancelot L. Farrar, Jr., *The Short-War Illusion: German Policy, Strategy and Domestic Affairs, August–December 1914* (Santa Barbara, Calif.: ABC-Clio, 1973); Stephen van Evera, "The Causes of War" (diss., University of California at Berkeley, 1984); and Jack Snyder, *The Ideology of the Offensive: Military Decision Making and the Disasters of 1914* (Ithaca: Cornell University Press, 1984), describe the evolution of the war plans of the several powers and the strategic assumptions on which they were predicated.

2. See Gerhard Ritter, *The Schlieffen Plan*, trans. Andrew and Eva Wilson (New York: Praeger, 1958); Farrar, *The Short-War Illusion*, pp. 1–33; and Snyder, *Ideology of the Offensive*, pp. 107–56.

3. Luigi Albertini, *The Origins of the War of 1914*, trans. Isabella M. Massey, 3 vols. (London: Oxford University Press, 1952), 3: 14–45, provides an authoritative account of German deliberations. However, controversy surrounds the matter of when German leaders received word of the Russian mobilization. For different points of view on this question and whether or not Germany acted prematurely, see Gerhard Ritter, *The Sword and the Scepter: The Problem of Militarism in Germany*, 4 vols., trans. Heinz Norden (Coral Gables: University of Miami Press, 1970), 2: 263–75, and Fritz Fischer, *War of Illusions: German Policies from 1911 to 1914*, trans. Marian Jackson (New York: Norton, 1975), pp. 488–501.

4. On Tannenburg see Edmund Ironside, *Tannenburg* (Edinburgh: Blackwood, 1933), and Norman Stone, *The Eastern Front, 1914–1917* (New York: Scribners, 1975), pp. 44–69. Alexander Solzhenitsyn's *August 1914*, trans. Michael Glenny (New York: Farrar, Straus & Giroux, 1971), provides an insightful and readable account of the causes of Russia's defeat.

5. Alexander von Kluck, *The March on Paris and the Battle of the Marne* (New York: Longmans, Green, 1920), is an excellent although not entirely impartial history by Germany's most capable commander on the Western front in 1914; also worthwhile is the account by his chief of staff, H. von Kuhl, *The Marne Campaign, 1914* (Ft. Leavenworth, Kans.: Command and General Staff School Press, 1936). For a more recent and well-balanced account, see Corelli Barnett, *The Swordbearers: Supreme Command in the First World War* (Bloomington: Indiana University Press, 1975).

6. This argument is elaborated in Richard Ned Lebow, "The Soviet Offensive in Europe: The Schlieffen Plan Revisited?" *International Security* 9 (Spring 1985), pp. 44–78. On reinforcement rates and force ratios at various stages of mobilization see Robert Lucas Fischer, *Defending the Central Front: The Balance of Forces*, Adelphi Paper no. 127 (London: International Institute of Strategic Studies, 1976), pp. 15–24; William P. Mako, *U.S. Ground Forces and the Defense of Central Europe* (Washington, D.C.: Brookings, 1983), pp. 52–58; and Fen Osler Hampson, "Groping for Technical Panaceas: The European Conventional Balance and Nuclear Stability," *International*

Security 8 (Winter 1983–84), pp. 64–69, which evaluates five different studies of mobilization under varying attack scenarios.

7. See Richard K. Betts, *Surprise Attack: Lessons for Defense Planning* (Washington, D.C.: Brookings, 1982), p. 203.

8. On the Soviet preference for conventional warfare see David M. Glantz, "Soviet Offensive Ground Doctrine since 1945," *Air University Review*, March–April 1983, pp. 25–35; Stephen M. Meyer, *Soviet Theatre Nuclear Forces*, Part I: *Development of Doctrine and Objectives*, Adelphi Paper no. 187 (London: International Institute of Strategic Studies, 1984); Phillip A. Petersen and John G. Hines, "The Conventional Offensive in Soviet Theater Strategy," *Orbis* 27 (Fall 1983), pp. 695–739; and Michael MccGwire, *Soviet Military Objectives in a World War* (Washington, D.C.: Brookings, forthcoming).

9. The United States denies that P-IIs are capable of reaching Moscow and insists that the Soviets would have twelve, not six, minutes of warning. Soviet concern about the strategic implications of NATO's modernization of intermediate nuclear forces is discussed by Strobe Talbott, *Endgame: The Inside Story of SALT II* (New York: Harper & Row, 1980), pp. 61 and 84; Raymond L. Garthoff, *Détente and Confrontation: American-Soviet Relations from Nixon to Reagan* (Washington, D.C.: Brookings, 1985), pp. 867, 869, 881–83; Stephen M. Meyer, *Soviet Theatre Nuclear Forces*, Part II: *Capabilities and Intentions*, Adelphi Paper no. 188 (London: International Institute of Strategic Studies, 1984), pp. 40–42; William V. Garner, *Soviet Threat Perceptions of NATO's Eurostrategic Missiles* (Paris: Atlantic Institute for International Affairs, 1983); and Graeme P. Auton, "European Security and the INF Dilemma: Is There a Better Way?" *Arms Control* 5 (May 1984), pp. 3–59. See also *Whence the Threat to Peace?* (Moscow: USSR Ministry of Defense, 1982), p. 60.

10. Desmond Ball, *Can Nuclear War Be Controlled?* Adelphi Paper no. 169 (London: International Institute of Strategic Studies, 1981), pp. 5–6, offers a good description of the force multiplier effect of strategic C^3I.

11. Bruce G. Blair, *Strategic Command and Control: Redefining the Nuclear Threat* (Washington, D.C.: Brookings, 1985), p. 285.

12. *Seminar on Command, Control, Communications, and Intelligence* (Cambridge: Center for Information Policy Research, Harvard University, 1982), p. 8.

13. William J. Broad, "Experts Say Satellite Can Detect Soviet War Steps," *New York Times*, 25 January 1985, p. A12, contains interviews with Paul Stares and Paul Bracken on the analytical methods applied to electronic intelligence to detect Soviet war preparations.

14. Good descriptions of the U.S. early warning system are provided by Congressional Budget Office, *Strategic Command, Control, and Communications: Alternative Approaches to Modernization* (Washington, D.C., 1981); Ball, *Can Nuclear War Be Controlled?* pp. 38–40; Paul Bracken, *The Command and Control of Nuclear Forces* (New Haven: Yale University Press, 1983), pp. 5–73; and Blair, *Strategic Command and Control*. The following narrative draws heavily upon these works.

15. The warning system is being upgraded. Phased-array radars will replace older, conventional radars in Greenland and at Fylingdales United Kingdom. The Thule radar will be dual faced, permitting it to scan 270 degrees, while the three-faced Fylingdales radar will scan a full 360 degrees.

16. Bracken, *Command and Control of Nuclear Forces*, pp. 21–24, provides an overview of the mechanisms for ensuring negative control.

17. The NCA has never been defined in unclassified documents. It is believed to consist of the president, the joint chiefs of staff, and/or the secretary of defense.

PALs were designed to prevent terrorists or other unauthorized individuals from gaining access to nuclear weapons that were deployed overseas. They also function as a check on the weapons' operators themselves. See U.S. Congress, House, Committee on International Relations, *First Use of Nuclear Weapons: Preserving Responsible Control*, 94th Cong., 2d sess. (1976), p. 93; Milton Leitenberg, "Background Materials on Tactical Nuclear Weapons Primarily in the European Context," in Stockholm International Peace Research Institute, ed., *Tactical Nuclear Weapons: European Perspectives* (London: Taylor & Francis, 1978), pp. 41–42.

18. See David A. Anderton, *Strategic Air Command* (New York: Scribners, 1977), pp. 16–17. Desmond Ball, in "Nuclear War at Sea," *International Security* 10 (Winter 1985–86), pp. 3–31, describes the launch procedure for U.S. SLBMs: "The submarine's captain and the executive officer each have one of two keys necessary to open a safe containing specific launch instructions. In the event the submarine received an emergency war message, the two men would, in the presence of a third officer, remove these instructions from the safe. The captain would then open the lock on a red "fire" button, to which only he has the combination. This action would start a carefully coordinated launch sequence involving at least 15 different individuals at various stations on the boat. To actually launch a missile, it takes four officers in different parts of the submarine to turn keys or throw switches. . . . If one of these officers fails—or refuses—to do his part, the missile cannot be fired."

19. So argue Ball, *Can Nuclear War Be Controlled?* pp. 41–42; Ashton B. Carter, "The Command and Control of Nuclear War," *Scientific American* 252 (January 1985), pp. 32–39; and Blair, *Strategic Command and Control*, pp. 264–72.

20. See Ball, *Can Nuclear War Be Controlled?* pp. 9–25, and Blair, *Strategic Command and Control*, pp. 204–7. For a detailed discussion of the effects of nuclear detonation on communications, see Office of Technology Assessment, *MX Missile Basing* (Washington, D.C., 1981), and Samuel Glasstone and Philip Dolan, eds., *The Effects of Nuclear Weapons*, 3d ed. (Washington, D.C.: U.S. Atomic Energy Commission, 1977).

21. Cf. Glasstone and Dolan, *Effects of Nuclear Weapons*, pp. 514–40; Ball, *Can Nuclear War Be Controlled?* pp. 10–12; and Blair, *Strategic Command and Control*, pp. 321–26.

22. For details, see Ball, *Can Nuclear War Be Controlled?* pp. 18–21; Blair, *Strategic Command and Control*, pp. 201–7.

23. See Kurt Gottfried and Richard Ned Lebow, "Anti-Satellite Weapons: Weighing the Risks," *Daedalus* 114 (Spring 1985), 1: 147–70. For a discussion of Soviet and American satellite and antisatellite programs, see Stephen M. Meyer, "Soviet Military Programmes and the 'New High Ground,'" *Survival*, September–October 1983, pp. 204–15, and Paul Stares, *The Militarization of Space: U.S. Policy, 1945–1984* (Ithaca: Cornell University Press, 1985).

24. On GWEN see Blair, *Strategic Command and Control*, pp. 253–55, and William M. Arkin, "Preparing for World War IV," *Bulletin of the Atomic Scientists* 41 (May 1985), pp. 6–7. For a more general description of command and control modernization programs, see Congressional Budget Office, *Strategic Command, Control, and Communications;* Jonathon B. Tucker, "Strategic Command and Control Vulnerabilities: Dangers and Remedies," *Orbis* 26 (Winter 1983), pp. 941–63; and Blair, *Strategic Command and Control*, pp. 241–80.

25. According to Blair, *Strategic Command and Control*, pp. 201–7.

26. See ibid., p. 253; Daniel Ford, *The Button: The Pentagon's Command and Control*

System—Does It Work? (New York: Simon & Schuster, 1985), pp. 213–14; and interview with Desmond Ball, 27 August 1985.

27. Bracken, *Command and Control of Nuclear Forces,* pp. 179–237, provides a detailed description of the U.S. command structure.

28. According to Brig. Gen. Raymond V. McMillan, NORAD's deputy chief of staff, "Cheyenne Mountain is survivable, and we're sure it will survive long enough for us to do our mission. But if the enemy wants to keep throwing SS-18s at it, sooner or later he's going to crumple it." Quoted in Bruce Gumble, "Air Force Upgrading Defenses at NORAD," *Defense Electronics* 17 (August 1985), pp. 86–108.

29. See Ball, *Can Nuclear War Be Controlled?* pp. 15–18; Blair, *Strategic Command and Control,* pp. 159–66, 262–63; and Strategic Air Command, Office of the Historian, *Development of Strategic Air Command, 1946–1976* (Offut Air Force Base, Neb., 1976), pp. 58–59.

30. According to Ford, *The Button,* p. 156, reporting on an interview with a SAC communications officer on the problems of very low frequency. On ERCS see Blair, *Strategic Command and Control,* pp. 107, 166–69, 184–87.

31. Ball, *Can Nuclear War Be Controlled?* pp. 23–26; Blair, *Strategic Command and Control,* pp. 269–72.

32. See Ball, *Can Nuclear War Be Controlled?* pp. 16–17, 24–26; Blair, *Strategic Command and Control,* pp. 156–58, 169–79, 198–201, 265–69.

33. Miller is quoted in House Committee on International Relations, *First Use of Nuclear Weapons,* pp. 47, 71. See also Ball, *Can Nuclear War Be Controlled?* passim, and John D. Steinbruner, "Nuclear Decapitation," *Foreign Policy* no. 45 (Winter 1981–82), pp. 16–28. According to Blair, *Strategic Command and Control,* p. 208, "an attack on our command system could drastically reduce or even block retaliation, and such an approach continues to offer the most attractive solution to the enemy planners' problem of limiting damage to the Soviet Union."

34. Described in Blair, *Strategic Command and Control,* pp. 108–11, 147–75, 261–62.

35. See Thomas Powers, "Choosing a Strategy for World War III," *Atlantic,* November 1982, pp. 62–110.

36. Robert Jervis, *Perception and Misperception in International Politics* (Princeton: Princeton University Press, 1976), pp. 193–95.

37. Cf. remarks of Gen. Brent Scowcroft, *MITRE Corporation National Security Issues Symposium* (New Bedford, Mass.: MITRE Corp., 1981), p. 95

38. For a description of the countervailing strategy, see Department of Defense, *Annual Report for Fiscal Year 1981* (Washington, D.C., 1980), pp. 5–7. See also the 16 September 1980 testimony of Secretary of Defense Harold Brown before U.S. Congress, Senate, *Hearing before the Committee on Foreign Relations of the U.S. Senate on Presidential Directive 59,* 96th Cong., 2d sess. (Washington, D.C., 1981), pp. 6–10, and Walter Slocombe, "The Countervailing Strategy," *International Security* 5 (Spring 1981), pp. 18–27. An excellent critique is provided by Robert Jervis, *The Illogic of American Nuclear Strategy* (Ithaca: Cornell University Press, 1984). For subsequent doctrinal developments under the Reagan administration, see Leon Sloss and Marc Dean Millot, "U.S. Nuclear Strategy in Evolution," *Strategic Review* 22 (Winter 1984), pp. 19–28, and Jeffrey Richelson, "PD 59, NSDD-13, and the Reagan Strategic Modernization Program," *Journal of Strategic Studies* 6 (June 1983), pp. 125–46.

39. See Robert P. Berman and John C. Baker, *Soviet Strategic Forces: Requirements and Responses* (Washington, D.C.: Brookings, 1982), pp. 64–65; Dan L. Strode, "The Soviet Union and Modernization of the U.S. ICBM Force," in Barry R. Schneider, Colin S. Gray, and Keith B. Payne, eds., *Missiles for the Nineties: ICBMs and Strategic Policy* (Boulder, Colo.: Westview, 1984), pp. 135–58; Desmond Ball, "The Soviet

Union and the Control of Nuclear War," *Soviet Union–Union Soviétique* 10, pts. 2–3 (1983), pp. 201–17; and William T. Lee, "Soviet Nuclear Targeting Strategy," in Desmond Ball and Jeffrey T. Richelson, eds., *Strategic Nuclear Targeting* (Ithaca: Cornell University Press, 1986), pp. 84–108. Stephen M. Meyer, "Soviet Perspectives on the Paths to Nuclear War," in Graham T. Allison, Albert Carnesale, and Joseph S. Nye, Jr., eds., *Hawks, Doves, and Owls: An Agenda for Avoiding Nuclear War* (New York: Norton, 1985), pp. 167–205, argues that the Soviets have become less reliant on quick launch procedures because of the greater survivability of their ICBM force and its command and control. Payne also acknowledges these improvements but argues that they have been offset by new U.S. deployments. I examine this question in the conclusion to this book.

40. Reported in letter from Deputy Secretary of Defense Roswell Gilpatric to Sen. Hubert Humphrey, 23 August 1961, in U.S. Congress, Senate, Committee on Foreign Relations, *Hearings on S. 2180*, 87th Cong., 1st sess. (Washington, D.C., 1961), pp. 110–11.

41. See Powers, "Choosing a Strategy."

42. Desmond Ball, for example, believes that "it would probably not be possible to isolate the Soviet (leadership) completely from the strategic forces or completely to impair the Soviet strategic intelligence flow." Quoted in Ford, *The Button*, p. 127.

43. See, for example, Benjamin S. Lambeth, "Selective Nuclear Options in American and Soviet Strategic Policy" (Santa Monica, Calif.: Rand, 1976), pp. 18–21; Desmond Ball, *Politics and Force Levels: The Strategic Missile Program of the Kennedy Administration* (Berkeley: University of California Press, 1980), and "U.S. Strategic Forces: How Would They Be Used?" *International Security* 7 (Winter 1982–83), pp. 31–60; and Powers, "Choosing a Strategy," pp. 82–110.

44. See, for example, Henry Trofimenko, "The Theology of Strategy," *Orbis* 21 (Fall 1977), pp. 497–515.

45. On LeMay's outspokenness see David A. Rosenberg, "A Smoking Radiating Ruin at the End of Two Hours: Documents on American Plans for Nuclear War with the Soviet Union, 1954–55," *International Security* 6 (Winter 1981–82), pp. 3–38; Richard Ned Lebow, "Windows of Opportunity: Do States Jump through Them?" *International Security* 9 (Summer 1984), pp. 147–86; Lawrence Freedman, *The Evolution of Nuclear Strategy* (New York: St. Martin's, 1981), p. 129; Fred Kaplan, *The Wizards of Armegeddon* (New York: Simon & Schuster, 1983), p. 133; and Curtis E. LeMay with Dale O. Smith, *America Is in Danger* (New York: Funk & Wagnalls, 1968), pp. 82–83.

46. U.S. Congress, Senate, Hearings before the Senate Committee on Armed Services, *MX Missile Basing System and Related Issues*, 98th Cong., 1st sess. (Washington, D.C., 1983), p. 417.

47. Other critics of U.S. policy have also expressed concern about apparent contradictions between declared and actual policy. See *Science*, 6 June 1986, p. 1187, for examples.

48. This point is also made by Richard K. Betts, "Surprise Attack and Preemption," in Allison, Carnesale, and Nye, *Hawks, Doves, and Owls*, pp. 54–79.

49. John Erikson, "The Soviet View of Deterrence: A General Survey," *Survival* 24 (November–December 1982), pp. 242–51, makes the point that the American decision not to pursue any serious civil defense program is also sometimes taken as proof of American intentions to launch a first strike.

50. Napoleon to Eugène, 30 April 1809, *Correspondance de Napoléon Ier.*, 32 vols. (Paris: Plon, 1858–70), 18: 525, no. 15144.

51. This phenomenon is analyzed by Ken Booth, *Strategy and Ethnocentrism* (New

York: Holmes & Meier, 1979), pp. 122–28, and Klaus Knorr, "Threat Prescriptions," in Knorr, ed., *Historical Dimensions of National Security Policy* (Leavenworth, Kans.: Allen, 1971), pp. 78–119.

52. Steinbruner, "Nuclear Decapitation," pp. 16–28.

53. Ibid., p. 19.

54. A bolt-from-the-blue strike, by contrast, would be much more likely to succeed in preventing retaliation. The adversary's strategic forces would be at day-to-day readiness, and more important, enough military and political officials in the chain of command would be sufficiently incredulous about what their sensors were telling them that they might respond too slowly. However, such an attack would not be preemption. It would be preventive war: the result of a cold-blooded decision to launch a surprise attack in circumstances where war could readily have been avoided by doing nothing. This seems the most unlikely kind of attack.

55. Interview with Desmond Ball, 27 August 1985. See also Jonathan B. Tucker, "Strategic Command and Control: America's Achilles Heel?" *Technology Review* 86 (August–September 1983), pp. 39–49. I treat this subject in greater detail in the next chapter.

56. Hippel quoted in William M. Arkin, "Sleight of Hand with Trident II," *Bulletin of the Atomic Scientists* 40 (December 1984), pp. 5–6; see also Joel S. Wit, "American SLBM: Counterforce Options and Strategic Implications," *Survival* 24 (July–August 1982), pp. 163–74.

57. For the best open-source description of the various SIOP options, see Desmond Ball, "Toward a Critique of Strategic Nuclear Targeting," and "The Development of the SIOP, 1960–83," both in Ball and Richelson, *Strategic Nuclear Targeting*, pp. 15–34, 57–83. See also Peter Pringle and William Arkin, *SIOP: The Secret U.S. Plan for Nuclear War* (New York: Norton, 1983).

58. So argues Bracken, p. 210, and before him Thornton Read, "Limited Strategic War and Tactical Nuclear War," in Klaus Knorr and Thornton Read, *Limited Strategic War* (New York: Praeger, 1962), pp. 105–7. Steinbruner, "Nuclear Decapitation," also stresses the risks inherent in a strategy of decapitation.

59. For an introduction to this literature, see C. Fritz and E. Marks, "The NORC Studies of Human Behavior in Disaster," *Journal of Social Issues* 10 (January 1954), pp. 26–41; Martha Wolfenstein, *Disaster* (Glencoe, Ill.: Free, 1957), pp. 11–30; and Irving L. Janis, "Psychological Effects of Warnings," in George Baker and Dwight Chapman, eds., *Man and Society in Disaster* (New York: Basic, 1962), pp. 55–92.

60. Interview with Amos Tversky, in William F. Allman, "Staying Alive in the 20th Century," *Science 85* 6 (October 1985), pp. 31–37. See also Daniel Kahneman, Paul Slovic, and Tversky, *Judgment under Uncertainty: Heuristics and Biases* (New York: Cambridge University Press, 1982), which reprints many pioneering studies by the coauthors and their colleagues in the field of cognitive bias and decision making. For more recent work that is particularly relevant to the propensity of people to gamble, see Tversky and Maya Bar-Hillel, "Risk: The Long and Short," *Journal of Experimental Psychology* 9 (October 1983), pp. 713–17; Tversky and Kahneman, "Extensional versus Intuitive Reasoning: The Conjunction Fallacy in Probability Judgment," *Psychological Review* 90 (October 1983), pp. 293–315; and Eric Johnson and Tversky, "Representatives of Perceptions of Risks," *Journal of Experimental Psychology: General* 113 (March 1984), pp. 55–70.

61. This is a central theme of Lebow, *Between Peace and War*, and Robert Jervis, Richard Ned Lebow, and Janice Gross Stein, *Psychology and Deterrence* (Baltimore: Johns Hopkins University Press, 1985).

62. War-winning arguments are put forward by Paul Nitze, "Assuring Strategic Stability in an Era of Detente," *Foreign Affairs* 54 (January 1976), pp. 207–32; Colin S. Gray, "Nuclear Strategy: A Case for a Theory of Victory," *International Security* 4 (Summer 1979), pp. 54–87; and Keith B. Payne, *Nuclear Deterrence in U.S.-Soviet Relations* (Boulder, Colo.: Westview, 1982). Jervis, *Illogic of American Nuclear Strategy*, pp. 57–63, offers a good critique.

63. Office of Technology Assessment, *The Effects of Nuclear War* (Washington, D.C., 1979), pp. 10, 94–106, estimates up to 160 million immediate U.S. deaths from a Soviet attack against a wide range of military and economic targets, with "tens of thousands" dying in the near-term future and "millions" in the longer term from radiation and the results of economic disruption. Arthur M. Katz, *Life after Nuclear War: The Economic and Social Impacts of Nuclear Attack on the United States* (Cambridge, Mass.: Ballinger, 1982), surmises that an attack of 300 to 500 warheads would result in U.S. casualties of 35–45 percent of the total population and 45–65 percent of the urban population. For estimates of damage to the Soviet Union, see Department of Defense, *Annual Report for Fiscal Year 1982* (Washington, D.C., 1981), pp. 37–38, 56–59, and Desmond Ball, *Targeting for Strategic Deterrence*, Adelphi Paper no. 185 (London: International Institute of Strategic Studies, 1983), pp. 19–21.

64. In addition to Katz, *Life after Nuclear War*, and OTA, *Effects of Nuclear War*, see Bruce W. Bennett, *Fatality Uncertainties in Limited Nuclear War* (Santa Monica, Calif.: Rand, 1977). Glasstone and Dolan's *Effects of Nuclear Weapons* is most pessimistic, because of the emphasis it places on the likely synergistic effects of disruption and destruction of agriculture, public health, medical care, and other critical components of modern life. As if this were not bad enough, David S. Greer of Brown University Medical School suggests that an AIDS epidemic could be expected to take its toll of the survivors of any nuclear war. *New York Times*, 22 September 1985, p. 27.

65. Interview with Michael MccGwire, 20 June 1985.

66. See Lebow, "Windows of Opportunity," pp. 155–57.

67. McGeorge Bundy, "To Cap the Volcano," *Foreign Affairs* 48 (October 1969), pp. 1–20.

68. Robert S. McNamara, *The Fiscal Year 1969–70 Defense Program and the 1969 Defense Budget* (Washington, D.C.: Department of Defense, 1969), pp. 50, 57; Alain C. Enthoven and K. Wayne Smith, *How Much Is Enough? Shaping the Defense Program, 1961–69* (New York: Harper & Row, 1971), pp. 207–8; and Frank von Hippel, "The Effects of Nuclear War," in David W. Hafemeister and Dietrich Schroeer, eds., *Physics, Technology and the Nuclear Arms Race* (New York: American Institute of Physics, 1983), calculate that these fatality levels could result from even lower levels of megatonnage.

69. Jervis, *Illogic of American Nuclear Strategy*. See also Richard Ned Lebow, "Misconceptions in American Strategic Assessment," *Political Science Quarterly* 97 (Summer 1982), pp. 187–206.

70. See Rosenberg, "A Smoking Radiating Ruin," for a description.

71. Lebow, "Misconceptions in American Strategic Assessment," makes this argument.

72. On the nuclear winter effect see R. P. Turco et al., "Nuclear Winter: Global Consequences of Multiple Nuclear Explosions," and Paul R. Ehrlich et al., "Long-Term Biological Consequences of Nuclear War," *Science*, 23 December 1983, pp. 1283–92, 1293–1300; Ehrlich et al., *The Cold and the Dark: The World after Nuclear War* (New York: Norton, 1984); and Carl Sagan, "Nuclear War and Climatic Catastro-

phe: Some Policy Implications," *Foreign Affairs* 62 (Winter 1983–84), pp. 257–92. For an early and predictable dissenting view, see Edward Teller, "Widespread After-Effects of Nuclear War," *Nature*, 23 August 1984, pp. 621–24. Caution about the uncertainty of nuclear winter effects is also urged by the National Research Council, Committee on the Atmospheric Effects of Nuclear Weapons, *The Effects on the Atmosphere of a Major Nuclear Exchange* (Washington, D.C.: National Academy Press, 1985), and an official study, Department of Defense, *The Potential Effects of Nuclear War on the Climate: A Report to the Congress* (Washington, D.C., March 1985).

73. On NDS, formerly known as IONDS, see Blair, *Strategic Command and Control*, pp. 261–64, 273.

74. On NEACP and Looking Glass vulnerability see Ball, *Can Nuclear War Be Controlled?* pp. 15–18; Blair, *Strategic Command and Control*, pp. 187–201.

75. Cf. Ball, *Can Nuclear War Be Controlled?*

76. Gottfried and Lebow, "Anti-Satellite Weapons," and Stares, *Militarization of Space*, discuss the vulnerability of satellites and their ground stations.

77. This point is made by Ball, *Targeting for Strategic Deterrence*, pp. 31–32, and Jervis, *Illogic of American Nuclear Strategy*, pp. 119–20.

78. Snyder, *Ideology of the Offensive*, pp. 125–56, documents the pessimism of Schlieffen and Moltke.

79. See, for example, senators Sam Nunn and Dewey F. Bartlett in U.S. Congress, Senate, Armed Services Committee, *NATO and the New Soviet Threat*, 95th Cong., 2d sess. (Washington, D.C., 1977); Phillip A. Karber, *The Impact of New Conventional Technologies on Military Doctrine and Organization in the Warsaw Pact*, Adelphi Paper no. 144 (London: International Institute of Strategic Studies, 1978); and European Security Study Group, *Strengthening Conventional Deterrence in Europe: Proposals for the 1980s* (New York: St. Martin's, 1983).

80. The most compelling presentation for this position is made by Barry R. Posen, "Inadvertent Nuclear War? Escalation and NATO's Northern Flank," *International Security* 7 (Fall 1982), pp. 28–54, who argues that nuclear escalation could be an unintended consequence of a decision to fight a conventional war. On the northern flank, conventional fighting could prompt nuclear preemption because (1) the offensive inclination of NATO naval forces would make them likely to strike at Soviet naval bases, destroying or putting in jeopardy important strategic assets; (2) attacks against either side's submarines and C^3I facilities might be interpreted as the prelude to strategic operations; (3) the fog of war invariably results in actions based on incomplete or inaccurate information which have unforseen consequences.

81. In Ball, "Nuclear War at Sea."

82. This situation in some ways resembles the classic prisoner's dilemma. Both sides would be better off if they exercised restraint but stand to lose most if they do while their adversary does not. The logic of the situation could lead one or both to attack even though both would prefer not to do so. This situation has been intensively studied, especially in repetitive encounters. In a one-time encounter, the situation that the superpowers would confront, prediction is impossible. See Anatol Rapoport and Albert M. Chammah, *Prisoner's Dilemma* (Ann Arbor: University of Michigan Press, 1965), and Robert Jervis, "Cooperation under the Security Dilemma," *World Politics* 30 (January 1978), pp. 167–214. See also Robert Axelrod's provocative study, *The Evolution of Cooperation* (New York: Basic, 1984).

83. This argument is most forcefully made by John Steinbruner, "National Security and the Concept of Strategic Stability," *Journal of Conflict Resolution* 22 (September 1978), pp. 411–28, and Bracken, *Command and Control of Nuclear Forces*.

84. See Chalmers A. Johnson, *An Instance of Treason: Ozaki Hotsumi and the Sorge Spy Ring* (Stanford: Stanford University Press, 1964), and Frederick W. Deakin, *The Case of Richard Sorge* (London: Chatto & Windus, 1966).

85. The nature and extent of U.S. advance warning about Pearl Harbor are discussed by Roberta Wohlstetter, *Pearl Harbor: Decision and Warning* (Stanford: Stanford University Press, 1962), and Gordon W. Prange, *At Dawn We Slept: The Untold Story of Pearl Harbor* (New York: Penguin, 1982), pp. 78–88.

86. For an example, see George F. Kennan's description of the Franco-German war scare of 1875 in *The Decline of Bismarck's European Order: Franco-Russian Relations, 1871–1890* (Princeton: Princeton University Press, 1979), pp. 11–26.

87. This theme is developed in Lebow, "Windows of Opportunity."

88. Douglas M. Hart, "Soviet Approaches to Crisis Management: The Military Dimension," *Survival* 26 (September–October 1984), pp. 214–22, gives a detailed description of the operations that likely would be part and parcel of a Soviet strategic alert.

89. Janice Gross Stein, "Calculation, Miscalculation, and Conventional Deterrence II: The View from Jerusalem," in Jervis, Lebow, and Stein, *Psychology and Deterrence*, pp. 60–88, offers an illuminating discussion of the difficulties of relying on tactical indicators of attack.

90. A. J. P. Taylor, "War by Time-Table," in *Purnell's History of the Twentieth Century* (New York: Purnell, 1974), 2: 442–48; Albertini, *Origins of the War of 1914*, 3: 253, offers the same judgment.

91. Lebow, *Between Peace and War*, pp. 254–63, and Jack S. Levy, "Organizational Routines and the Causes of War," forthcoming in *International Studies Quarterly*, explore the role of these expectations upon crisis decision making in 1914.

92. See Betts, *Surprise Attack*, p. 203.

93. DefCon I is equivalent to war, DefCon II signifies that attack is imminent, and DefCon III increases the readiness of U.S. forces without any judgment as to the likelihood of war. American forces are normally kept at DefCon IV or V. They have been brought up to DefCon III on a worldwide basis on only four occasions: in mid-1960 during Eisenhower's trip to Paris, during the Cuban missile crisis of 1962, following Kennedy's assassination in November 1963, and in 1973, in the aftermath of the Soviet threat to intervene in the Middle East. American forces in the Pacific were also put on DefCon III in 1968 in response to events in Indochina. SAC was brought up to DefCon II for the one and only time during the Cuban crisis. On the 1962 and 1973 alerts see U.S. Department of Defense, *Annual Report of the Secretary of Defense for 1963* (Washington, D.C., 1963), p. 5, and Scott Sagan, "Nuclear Alerts and Crisis Management," *International Security* 9 (Spring 1985), pp. 99–139. On 1973 see also Henry Kissinger, *Years of Upheaval* (Boston: Little, Brown, 1982), pp. 575–90; Barry M. Blechman and Douglas M. Hart, "The Political Utility of Nuclear Weapons: The 1973 Middle East Crisis," *International Security* 7 (Summer 1982), pp. 132–56; and Garthoff, *Détente and Confrontation*, pp. 374–84.

94. So argues Kissinger, *Years of Upheaval*, p. 588.

95. In the past this could be attributed entirely to technical reasons. *Air Force Magazine*, March 1978, p. 51; U.S. Department of Defense, *Soviet Military Power* (Washington, D.C., 1981), pp. 6–7, 55; and an interview with Defense Secretary Harold Brown, "Could Russia Blunder into Nuclear War?" *U.S. News and World Report*, 5 September 1977, p. 18, reveals that Soviet ICBMs could be quickly brought up to full alert but could not be held there for long because of the limited endurance of the gyroscopes essential for their guidance. The Central Intelligence Agency,

"Major Consequences of Certain US Courses of Action in Cuba," SNIE 11-19-62, declassified (20 October 1962), p. 2, determined that Soviet regional forces, then SS-4 and SS-5s, required eight hours' preparation before they could be launched and could be kept at that level of readiness only for five hours. Several years after Cuba, nuclear weapons were placed on day-to-day-alert ICBMs, but these constituted only a small percentage of the overall Soviet missile force. Department of Defense, *Annual Report for Fiscal Year 1979* (Washington, D.C., 1978), p. 72. Today the situation has from the Soviet point of view improved considerably. It is generally assumed that about 20 to 25 percent of the older SS-9s, 11s, and 13s are on day-to-day alert whereas about 90 percent of the SS-18s and 19s can be maintained at that status. Berman and Baker, *Soviet Strategic Forces,* p. 86, report that storable liquid fuels permit the most recent generation of Soviet ICBMs to be launched in four to eight minutes. Soviet long-range bomber aircraft are never kept on standby alert, although they do engage in periodic practice alerts and Arctic staging exercises; see Joseph J. Kruzel, "Military Alerts and Diplomatic Signals," in Ellen P. Stern, ed., *The Limits of Military Intervention* (Beverly Hills: Sage, 1977), pp. 83–100. The first Soviet SSBNs, the Yankee class, began continuous patrolling along the Atlantic coast in 1969, as William Beecher reported in the *New York Times,* 24 April 1970, p. 6. In 1974, according to Berman and Baker, *Soviet Strategic Forces,* p. 18, Soviet Delta-class SSBNs entered service, and several are regularly on station, mostly in the vicinity of the Soviet Union. They complement the several forward-deployed Yankee-class vessels off the U.S. Atlantic and Pacific coasts. U.S. Congress, House, Hearings before the Subcommittee of the Appropriations Committee, *Department of Defense Appropriations for 1980,* pt. III, 96th Cong., 1st sess., 1979, p. 476, reports on-station rates for Soviet SSBNs as low as 10 to 15 percent. Meyer, "Soviet Perspectives on the Paths to Nuclear War," n. 26, ventures the judgment that the SSBN alert rate is slowly rising and today is probably in the range of 20 to 25 percent.

96. On Soviet expectations about how war would arise see Meyer, "Soviet Perspectives on the Paths to Nuclear War," p. 178.

97. Richard E. Neustadt and Graham T. Allison, "Afterword," in Robert F. Kennedy, *Thirteen Days: A Memoir of the Cuban Missile Crisis* (New York: Norton, 1971), p. 113, is the only source that makes any reference to the Soviet Union raising the alert level of its nuclear forces during the Cuban crisis. Other studies of the crisis deny that this was the case.

98. See *U.S. News and World Report,* 24 December 1973. Hart, "Soviet Approaches to Crisis Management," states that the Soviet Mediterranean flotilla was increased to 96 ships, including 34 major surface combatants and 23 submarines. On 26 October it began anticarrier exercises that continued nonstop for eight days. Kissinger, *Years of Upheaval,* p. 584, confirms that Soviet naval activity in the Mediterranean surged during the crisis to reach an all-time high, by his count, of one hundred naval vessels.

99. For example, Hart, "Soviet Approaches to Crisis Management," p. 221.

100. Although Soviet ICBMs on day-to-day alert can be launched within minutes, it is likely that any decision to launch would be preceded by efforts to bring all Soviet forces, strategic and conventional, up to a higher state of readiness. Some of the steps these efforts would entail are described by Meyer, "Soviet Perspectives on the Paths to Nuclear War," pp. 181–82, and in greater detail by Hart, "Soviet Approaches to Crisis Management," pp. 117–20. Alerts of this kind would be detected by the United States and would provide American policy makers with time in which to consider their response.

101. Lebow, *Between Peace and War*, pp. 238–42, reviews the evidence for the thesis that misleading intelligence played an important role in the Russian, German, and French mobilizations in 1914.

102. Ulrich Trumpener, "War Premeditated? German Intelligence Operations in July 1914," *Journal of Central European History* 9 (March 1976), pp. 58–85, assesses the operations of German military intelligence and their influence on German decision making in the crisis; L. C. F. Turner, "Russian Mobilization in 1914," *Journal of Contemporary History* 3 (January 1968), pp. 65–88, portrays the generals more sympathetically, arguing that premobilization in Russia really did constitute a form of early mobilization.

103. The case for this is made by Hart, "Soviet Approaches to Crisis Management," p. 217, and by John Dziak, Richard Heuer, and William van Cleave in a forthcoming study, *Soviet Strategic Deception*, ed. Brian Dailey and Patrick Parker.

104. *New York Times*, 3 December 1980, p. 1; 8 December 1980, p. 1.

105. Quoted in *New York Times*, 2 January 1981, p. 3.

106. *New York Times*, 9 January 1981, p. 4.

107. Interviews with involved officials indicate that American intelligence had detected the Soviet buildup along the Afghan border, but most officials had refused to believe it was a prelude to invasion. Later, the commitment to avoid making the same mistake a second time made the intelligence community even more committed to the expectation that an invasion of Poland was imminent.

108. Stephen S. Kaplan et al., *The Diplomacy of Power: Soviet Armed Forces as a Political Instrument* (Washington, D.C.: Brookings, 1981), p. 97.

109. Barry M. Blechman and Stephen M. Kaplan, *Force without War: U.S. Armed Forces as a Political Instrument* (Washington, D.C.: Brookings, 1978), p. 23.

110. Ibid., p. 41.

111. This is the central conclusion of Richard K. Betts, *Soldiers, Statesmen, and Cold War Crises* (Cambridge: Harvard University Press, 1977).

112. Such an exercise is described by Pringle and Arkin in *SIOP: The Secret U.S. Plan for Nuclear War*, pp. 22–36. Codenamed "Ivy League" (was George Bush standing in for the president?), it was reported to be the largest such exercise held in thirty years and involved numerous high-ranking civilian and military officials.

113. See, for example, Gary D. Brewer and Martin Shubik, *The War Game: A Critique of Military Problem Solving* (Cambridge: Harvard University Press, 1979).

114. See Alexander L. George and Richard Smoke, *Deterrence in American Foreign Policy: Theory and Practice* (New York: Columbia University Press, 1974), pp. 550–61; Stephen Maxwell, *Rationality in Deterrence*, Adelphi Paper no. 50 (London: International Institute of Strategic Studies, 1968); Robert Jervis, "Deterrence Theory Revisited," *World Politics* 31 (January 1979), pp. 289–324; and Richard Ned Lebow, "Conclusions," to Jervis, Lebow and Stein, *Psychology and Deterrence*, pp. 203–32.

3. *Loss of Control*

1. See, for example, Elie Abel, *The Missile Crisis* (Philadelphia: Lippincott, 1966); Arthur M. Schlesinger, Jr., *A Thousand Days: John F. Kennedy in the White House* (Boston: Houghton Mifflin, 1965); Graham T. Allison, *Essence of Decision: Explaining the Cuban Missile Crisis* (Boston: Little, Brown, 1971); and Scott D. Sagan, "Nuclear Alerts and Crisis Management," *International Security* 9 (Spring 1985), pp. 99–139.

2. Robert F. Kennedy, *Thirteen Days: A Memoir of the Cuban Crisis* (New York: Norton, 1969), p. 62.

3. Bruce G. Blair, *Strategic Command and Control: Redefining the Nuclear Threat* (Washington, D.C.: Brookings, 1985).

4. Paul Bracken, *The Command and Control of Nuclear Forces* (New Haven: Yale University Press, 1983), p. 3.

5. Ibid.

6. Ibid., pp. 5–48, 179–219.

7. For example, Richard Halloran, "Military's Message System Is Overloaded, Officers Say," *New York Times*, 25 November 1985, p. 17.

8. Reported by Gary Hart and Barry Goldwater, *Recent False Alerts from the Nation's Missile Attack Warning System*, Report to the Senate Committee on Armed Services (Washington, D.C., 1980), p. 4; and *New York Times*, 11 November 1979, p. 30; 16 December 1979, p. 25; 18 June 1980, p. A16; 23 June 1980, p. 58.

9. Charles Perrow, *Normal Accidents: Living with High Risk Technologies* (New York: Basic, 1984), explores this theme at length and provides illustrations from complex organizations as diverse as nuclear power and petrochemical plants, aircraft and airways, and marine transportation.

10. Bracken, *Command and Control of Nuclear Forces*, p. 56.

11. See John Steinbruner, "Nuclear Decapitation," *Foreign Policy* no. 45 (Winter 1981–82), pp. 16–28.

12. See Congressional Research Service, *Authority to Order the Use of Nuclear Weapons* (Washington, D.C., 1975), and U.S. Congress, House, Subcommittee on International Security and Scientific Affairs, *Nuclear Weapons: Preserving Responsible Control*, 94th Cong., 2d sess. (Washington, D.C., 1976).

13. *U.S. News and World Report*, 5 October 1964, cited in Bracken, p. 198.

14. Ibid.

15. *New York Times*, 4 November 1977, p. A9.

16. U.S. Congress, *Nuclear Weapons*, indicates that at some time the commanders-in-chief of NORAD and the Pacific, Atlantic, and European commands have all had some kind of release authority.

17. Cf. Raymond Tate, "Worldwide C³I and Telecommunications," in *Seminar on Command, Control, Communications, and Intelligence* (Cambridge: Program on Information Resources Policy, Harvard University, 1980), p. 43.

18. Bracken, *Command and Control of Nuclear Forces*, pp. 165–69 and 199–200.

19. American field commanders I have talked to estimate that it would take a minimum of twenty-four hours for a request to use a small number of nuclear weapons against advancing Warsaw Pact forces to work its way up through the various American layers of command and NATO committees.

20. See Phil Stanford, "Who Pushes the Button?" *Parade*, 28 March 1976, and Desmond Ball, "Nuclear War at Sea," *International Security* 10 (Winter 1985–86), pp. 3–31.

21. *Los Angeles Times*, 14 October 1984, p. 28, and Ball, "Nuclear War at Sea."

22. See Robert P. Berman and John C. Baker, *Soviet Strategic Forces: Requirements and Responses* (Washington, D.C.: Brookings, 1982), pp. 58, 62–65. For a critical evaluation of the thesis that these SSBNs are a reserve force, see Jan S. Breemer, "The Soviet Navy's SSBN Bastions: Evidence, Inference, and Alternative Scenarios," *Journal of the Royal United Services Institute* 130 (March 1985), pp. 18–26.

23. Adm. James D. Watkins, USN, "The Maritime Strategy," in *The Maritime Strategy*, supplement to *Proceedings of the U.S. Naval Institute* (January 1986), p. 13.

24. Ball, "Nuclear War at Sea," describes Holystone missions in which U.S. submarines penetrate Soviet territorial waters and even naval bases in order to

perform a variety of missions: plug into underwater communications cables, monitor Soviets SLBM tests, record "voice autographs" of Soviet submarines, or gather other kinds of intelligence. *New York Times*, 25 May 1975, p. 42; 6 July 1975, pp. 1 and 26; and 20 January 1976, pp. 1 and 4, reported collisions in Soviet waters between U.S. submarines on Holystone missions and Soviet naval vessels.

25. Worth H. Bagley, *Sea Power and Western Security: The Next Decade*, Adelphi Paper no. 139 (London: International Institute of Strategic Studies, 1977), p. 12.

26. See Ball, "Nuclear War at Sea," on this question.

27. Bracken, *Command and Control of Nuclear Forces*, pp. 59–60.

28. Ibid., p. 64.

29. The pioneering study of Fritz Heider, *The Psychology of Interpersonal Relations* (New York: Wiley, 1958), demonstrated the prevalence of causal schemas in the perception of social relations. Albert E. Michotte, *The Perception of Causality* (New York: Basic, 1963), showed the extent to which people perceive events in causal sequences even when they know that the relation between events is incidental. Additional important work on causal attribution has been done by E. E. Jones et al., *Attribution: Perceiving the Causes of Behavior* (Morristown, N.J.: General Learning, 1971), and Lee Ross, "The Intuitive Psychologist and His Shortcomings: Distortions in the Attribution Process," in Leonard Berkowitz, ed., *Cognitive Theories in Social Psychology* (New York: Academic, 1978), pp. 173–200. Robert Jervis, *Perception and Misperception in International Politics* (Princeton: Princeton University Press, 1976), pp. 319–42, discusses many of the implications of this research for international relations.

30. Bracken, *Command and Control of Nuclear Forces*, pp. 65–68, gives an example of how this happened in 1956.

31. See President's Commission on the Accident at Three Mile Island, *The Need for Change: The Legacy of TMI* (Washington, D.C., 1979); Perrow, *Normal Accidents*, pp. 15–31, 137–41; and Paul C. Stern and Elliot Aronson, eds., *Energy Use: The Human Dimension* (New York: Freeman, 1984), pp. 4–5.

32. Perrow, *Normal Accidents*, p. 4.

33. Ibid., pp. 4–5.

34. Quoted in Hilliard Roderick with Ulla Magnusson, eds., *Avoiding Inadvertent War: Crisis Management* (Austin: Lyndon B. Johnson School of Public Affairs, University of Texas, 1983), p. 55.

35. For example, General Accounting Office, *NORAD's Missile Warning System: What Went Wrong?* (Washington, D.C., 1981), and U.S. Congress, House, Hearings before a Subcommittee of the Committee on Government Operations, *Failures of the North American Aerospace Defense Command's Attack Warning System*, 97th Cong., 1st sess. (Washington, D.C., 1981). Daniel Ford, *The Button: The Pentagon's Strategic Command and Control System* (New York: Simon & Schuster, 1985), pp. 78–84, summarizes some of these findings.

36. On this subject see Herbert Lin, "The Development of Software for Ballistic Missile Defense," *Scientific American* 253 (December 1985), pp. 46–53.

37. See Nicholas L. Johnson, *The Soviet Year in Space: 1983* (Colorado Springs: Brown Teledyne Engineering, 1984), pp. 29–30. Desmond Ball, "The Soviet Strategic Command, Control, Communications and Intelligence (C^3I) System," to appear in *C^3I Handbook* (Palo Alto, Calif.: EW Communications, 1986), provides a good overview of the architecture and capability of Soviet early warning.

38. The Soviet Union will soon have six operational Pechora-class large phased-array radars to supplement the older and less sophisticated network of eleven

Henhouse radars and two over-the-horizon radars. See Department of Defense, *Soviet Military Power* (Washington, D.C., 1984), pp. 20, 24.

39. See Zhores A. Medvedev, *Nuclear Disaster in the Urals,* trans. George Saunders (New York: Norton, 1979); J. R. Trabalka et al., *Analysis of the 1957–58 Soviet Nuclear Accident* (Oak Ridge, Tenn.: Oak Ridge National Laboratory, Report ORNL-5613, 1979); and Leslie Gelb, "Keeping an Eye on Russia," *New York,* 29 November 1981, pp. 31ff. Medvedev's assertion that a nuclear accident occurred was challenged by Los Alamos physicists William Stratton, Denny Stillman, Sumner Barr, and Harold M. Agnew in "Are Portions of the Urals Really Contaminated?" *Science* 206 (26 October 1979), pp. 423–25, who asserted that the radiation was more likely the result of fallout from a particularly dirty bomb test.

40. Stephen M. Meyer, "Soviet Perspectives on the Paths to Nuclear War," in Graham T. Allison, Albert Carnesale, and Joseph S. Nye, Jr., eds., *Hawks, Doves, and Owls: An Agenda for Avoiding Nuclear War* (New York: Norton, 1985), pp. 167–205, describes some examples.

41. See Berman and Baker, *Soviet Strategic Forces,* pp. 36–37, and Meyer, "Soviet Perspectives on the Paths to Nuclear War," pp. 175, 187–92. On readiness see Chapter 1, note 99.

42. See K. Vershinin, "The Influence of Scientific Technical Progress on the Development of the Air Force and Its Strategy in the Post War Period," and "The Development of the Operational Art of the Air Force," *Military Thought* no. 5 (1966), pp. 36–44, and no. 6 (1967), pp. 1–13, cited in Meyer, "Soviet Perspectives on the Paths to Nuclear War," p. 188 n. 48.

43. See John Barron, *KGB: The Secret Work of Soviet Secret Agents* (Pleasantville, N.Y.: Reader's Digest, 1974), p. 15; Oleg Penkovskiy, *The Penskovskiy Papers* (Garden City, N.Y.: Doubleday, 1965), p. 331; and Department of Defense, *Annual Report of the Secretary of Defense for Fiscal Year 1979* (Washington, D.C., 1978), p. 69.

44. Berman and Baker, *Soviet Strategic Forces,* p. 18.

45. So argues Stephen M. Meyer, "Space and Soviet Military Planning," in William Durch, ed., *National Interests and the Military Use of Space* (Cambridge, Mass.: Ballinger, 1984), pp. 61–68, and "Soviet Perspectives on the Paths to Nuclear War," p. 191.

46. Among them are Andrew Cockburn, *The Threat: Inside the Soviet Military Machine* (New York: Random House, 1983), p. 189, and Meyer, "Soviet Perspectives on the Paths to Nuclear War," p. 188, citing Aleksandr Prokhanov, "Yadernyy Shchit," *Literaturnaya Gazeta* no. 46 (17 November 1982), p. 10.

47. Michael McCGwire believes that Soviet leaders place so much importance on central control that they would delegate launch authority only at the point in a crisis or conventional war when they actually expected these weapons to be used. Interview, 5 November 1985.

48. Meyer, "Soviets Perspectives on the Paths to Nuclear War," pp. 190–91.

49. Outlined in Jack Snyder, *The Ideology of the Offensive: Military Decision Making and the Disasters of 1914* (Ithaca: Cornell University Press, 1984), pp. 142–45. Maneuvers were also orchestrated to guarantee that the kaiser's side always emerged victorious, a practice that the younger Moltke insisted had to end if he was to serve as commander-in-chief of the general staff. Corelli Barnett, *The Swordbearers: Supreme Command in the First World War* (Bloomington: Indiana University Press, 1975), p. 34.

50. I am indebted to Henry Kendall for this concept.

51. Perrow, *Normal Accidents,* pp. 50–54, 57–61. What is remarkable, according to

Perrow, is the continuing reluctance of the NRC and the industry to acknowledge the legitimacy of critics' concerns.

52. On the Soviet threat and American response see Alexander L. George, "The Arab-Israeli War of October 1973: Origins and Impact," in George, *Managing U.S.-Soviet Rivalry: Problems of Crisis Prevention* (Boulder, Colo.: Westview, 1979), pp. 139–54; Barry M. Blechman and Douglas M. Hart, "The Political Utility of Nuclear Weapons: The 1973 Middle East Crisis," *International Security* 7 (Summer 1982), pp. 132–56; Raymond L. Garthoff, *Détente and Confrontation: American-Soviet Relations from Nixon to Reagan* (Washington, D.C.: Brookings, 1985), pp. 374–84; Sagan, "Nuclear Alerts and Crisis Management"; and Henry Kissinger, *Years of Upheaval* (Boston: Little, Brown, 1982), pp. 575–90.

53. On British strategic forces and doctrine see John Baylis, *British Defence in a Changing World* (London: Croom Helm, 1977); Lawrence Freedman, *Britain and Nuclear Weapons* (London: Macmillan, 1980); Leon V. Sigal, "No First Use and NATO's Nuclear Policy," in John D. Steinbruner and Sigal, eds., *Alliance Security: NATO and the No-First Use Question* (Washington, D.C.: Brookings, 1983), pp. 106–33; Peter Malone, *The British Nuclear Deterrent* (New York: St. Martin's, 1984); and Christopher J. Bowie and Alan Platt, *British Nuclear Policymaking* (Santa Monica, Calif.: Rand, 1984). British and French strategic forces, their organization and basing, are described by William M. Arkin and Richard W. Fieldhouse, *Nuclear Battlefields* (Cambridge, Mass.: Ballinger, 1985).

54. On French nuclear capability and doctrine see Sigal, "No First Use and NATO's Nuclear Posture"; Robbin F. Laird, "The French Strategic Dilemma," *Orbis* 28 (Summer 1984), pp. 307–28, and *French Nuclear Forces in the 1980s and 1990s* (Alexandria, Va.: Center for Naval Analysis, 1983); and David S. Yost, *France's Deterrent Posture and Security in Europe*, pt. I: *Capabilities and Doctrine*, Adelphi Paper no. 194 (London: International Institute of Strategic Studies, 1984–85), and "French Nuclear Targeting," in Desmond Ball and Jeffrey Richelson, eds., *Strategic Nuclear Targeting* (Ithaca: Cornell University Press, 1986), pp. 127–56.

55. See Michael M. Harrison, *The Reluctant Ally: France and Atlantic Security* (Baltimore: Johns Hopkins University Press, 1981), and David S. Yost, *France's Deterrent Posture and Security in Europe*, pt. II: *Strategic and Arms Control Implications*, Adelphi Paper no. 195 (London: International Institute of Strategic Studies, 1984–85), pp. 21–34.

56. See Gregory Treverton, "China's Nuclear Forces and the Stability of Soviet-American Deterrence," in *The Future of Strategic Deterrence*, pt. I: *Papers from the IISS 21st Annual Conference*, Adelphi Paper no. 160 (London: International Institute of Strategic Studies, 1980), pp. 38–44; Gerald Segal, "China's Nuclear Posture for the 1980s," *Survival* 23 (January–February 1981), pp. 11–18, and *Defending China* (Oxford: Oxford University Press, 1985); and Robert S. Wang, "China's Evolving Strategic Doctrine," *Asian Survey* 24 (October 1984), pp. 1040–55.

4. *Miscalculated Escalation*

1. "The State of the Union," Address Delivered before a Joint Session of the Congress, 23 January 1980, in *Public Papers of the Presidents of the United States: Jimmy Carter, 1980–1981* (Washington, D.C., 1981), 1: 194–200.

2. Khrushchev made this threat to visiting American businessman William Knox, who was summoned to the Kremlin on 24 October for a three-hour interview. W. E.

Knox, "Close Up of Khrushchev during a Crisis," *New York Times Magazine*, 18 November 1962, pp. 32, 128–29. On the boarding of the *Marcula* see Graham Allison, *Essence of Decision: Explaining the Cuban Missile Crisis* (Boston: Little, Brown, 1971), pp. 35, 129–30.

3. See Lord Grey of Fallodon, *Twenty-Five Years: 1892–1916* (London: Hodder & Stoughton, 1925), pp. 1–18; Luigi Albertini, *The Origins of the War of 1914*, trans. Isabella M. Massey, 3 vols. (Oxford: Oxford University Press, 1952), 3: 380–86; and Michael G. Eckstein and Zara Steiner, "The Sarajevo Crisis," in F. Hinsley, ed., *British Foreign Policy under Sir Edward Grey* (London: Cambridge University Press, 1977), pp. 397–410.

4. W. Philip Davison, *The Berlin Blockade: A Study in Cold War Politics* (Princeton: Princeton University Press, 1958), pp. 154–55, 198–99.

5. Thomas C. Schelling, *Arms and Influence* (New Haven: Yale University Press, 1966), pp. 99–105.

6. For documentation, see Chapter 2, note 63.

7. See Richard Ned Lebow, *Between Peace and War: The Nature of International Crisis* (Baltimore: Johns Hopkins University Press, 1981), and Robert Jervis, Lebow, and Janice Gross Stein, *Psychology and Deterrence* (Baltimore: Johns Hopkins University Press, 1985), chaps. 3–4, 9.

8. See Neville Maxwell, *India's China War* (Garden City, N.Y.: Doubleday, 1972); Allen Whiting, *The Chinese Calculus of Deterrence: India and Indochina* (Ann Arbor: University of Michigan Press, 1975); and Lebow, *Between Peace and War*, pp. 148–228.

9. See Allen S. Whiting, *China Crosses the Yalu: The Decision to Enter the Korean War* (New York: Macmillan, 1960); Alexander L. George and Richard Smoke, *Deterrence in American Foreign Policy: Theory and Practice* (New York: Columbia University Press, 1974), pp. 184–234; and Lebow, *Between Peace and War*, pp. 148–228.

10. Nadav Safran, *From War to War: The Arab-Israeli Confrontation, 1948–1967* (New York: Pegasus, 1969), pp. 271–302, and Walter Laqueur, *The Road to War: The Origin and Aftermath of the Arab-Israeli Conflict of 1967–68* (Harmondsworth: Penguin, 1969), pp. 122–25, 254–72.

11. Leslie Dewart, "The Cuban Crisis Revisited," *Studies on the Left* 5 (Spring 1965), pp. 22–40, claims that Kennedy "sandbagged" Khrushchev: he encouraged him to believe that the United States would tolerate a Soviet military buildup in Cuba and then turned on him once that buildup was under way. Dewart's analysis, while original, is not based on any documentary evidence.

12. Washington's refusal to make a commitment to South Korea reflected unwillingness to confront a hard choice. The Truman administration wanted to discourage Soviet and North Korean aggression but also feared the prospect of being dragged into a land war on the Asian mainland. Dean Acheson also worried that a firm commitment to South Korea might encourage the Syngman Rhee regime to provoke a war with North Korea. Harry S Truman, *Memoirs* (Garden City, N.Y.: Doubleday, 1955), 2: 355–56. The evidence on this point is summarized by George and Smoke, *Deterrence in American Foreign Policy*, pp. 184–234, in their case study of the origins of the Korean conflict. A more recent account that makes use of additional primary source materials is David S. McLellan, *Dean Acheson: The State Department Years* (New York: Dodd, Mead, 1976), pp. 267–70.

13. Albertini, *Origins of the War of 1914*, 2: 579. Also see his discussion of this problem in 2: 479–85.

14. See ibid., 2: 479–85; L. C. F. Turner, *Origins of the First World War* (New York: Norton, 1970), pp. 109–10; Lebow, *Between Peace and War*, pp. 234–37; Jack Snyder, *The Ideology of the Offensive: Military Decision Making and the Disasters of 1914* (Ithaca: Cornell University Press, 1984), p. 185; and Jack S. Levy, "Organizational Routines and the Causes of War," forthcoming in *International Studies Quarterly*.

15. Kurt Lewin, *Field Theory in Social Science*, ed. D. Cartwright (New York: Harper & Row, 1964). See also Charles A. Kiesler, ed., *The Psychology of Commitment* (New York: Academic, 1971).

16. This evidence is reviewed in Robert Jervis, *Perception and Misperception in International Politics* (Princeton: Princeton University Press, 1976), pp. 383–406, and Irving L. Janis and Leon Mann, *Decision Making: A Psychological Analysis of Conflict, Choice and Commitment* (New York: Free Press, 1977), pp. 309–38.

17. This was particularly pronounced in Austria and Russia. See Albertini, *Origins of the War of 1914*, 2: 528–81, 651–86, and Lebow, *Between Peace and War*, pp. 136–37. On the public reaction see Barbara W. Tuchman, *The Guns of August* (New York: Macmillan, 1962), pp. 93–160.

18. See S. C. B. Lieven, *Russia and the Origins of the First World War* (New York: St. Martin's, 1983), p. 145.

19. See Albertini, *Origins of the War of 1914*, 2: 528–81, and L. C. F. Turner, "The Russian Mobilization in 1914," *Journal of Contemporary History* 3 (January 1968), pp. 65–88, for the background to the Russian mobilization. Sergei Sazonov, *Fateful Years, 1909–1916: The Reminiscences of Sergei Sazonov* (London: Stokes, 1928), p. 200, describes his own understanding of the meaning of mobilization at the time.

20. Max von Montgelas and Walter Schüking, eds., *Die deutschen Dokumente zum Kriegsausbruch 1914*, 3 vols. (Berlin: Deutsche Verlagsgesellschaft für Politik und Geschichte, 1922), 3: 487.

21. Ibid., p. 546.

22. Sidney Bradshaw Fay, *The Origins of the World War*, 2 vols. (New York: Macmillan, 1928), 2: 479–80.

23. Grey, *Twenty-Five Years*, pp. 330–31; Albertini, *Origins of the War of 1914*, 2: 480, 517–19.

24. G. P. Gooch and Harold Temperley, eds., *British Documents on the Origins of the War, 1898–1914*, 11 vols. (London: HMSO, 1926–28), 11: 146.

25. Ibid., 11: 185.

26. *Deutschen Dokumente zum Kriegsausbruch*, 1: 219.

27. Friedrich graf Pourtalès, *Am Scheidewege zwischen Krieg und Frieden: Meine letzten Verhandlungen in Petersburg, Ende Juli 1914* (Berlin: Deutsche Verlagsgesellschaft für Politik und Geschichte, 1919), p. 27. On Pourtalès's distortions and misrepresentations during the crisis and subsequently, in his memoirs, see Lebow, *Between Peace and War*, pp. 127–29.

28. *Deutsche Dokumente zum Kriesgausbruch*, 2: 342.

29. Pourtalès, *Am Scheidewege zwischen Krieg und Frieden*, p. 43, and Baron von Schilling, *How the War Began in 1914: Diary of the Russian Foreign Office from the 3rd to the 20th (Old Style) July, 1914*, trans. W. C. Bridge (London: Allen & Unwin, 1925), pp. 48–49.

30. Albertini, *Origins of the War of 1914*, 2: 539–81, and Turner, "Russian Mobilization of 1914," offer the best treatments of the deliberations leading up to the Russian decision to mobilize.

31. On this subject see Ernest R. May's illuminating "Cabinet, Tsar, Kaiser: Three

Approaches to Assessment," in May, ed., *Knowing One's Enemies* (Princeton: Princeton University Press, 1984), pp. 11–36, and William C. Fuller, "The Russian Empire," pp. 98–126, in the same volume. Barry R. Posen, *The Sources of Military Doctrine: France, Britain, and Germany between the World Wars* (Ithaca: Cornell University Press, 1984), offers an insightful theoretical treatment of this problem.

32. See David B. Ralston, *The Army of the Republic: The Place of the Military in the Political Evolution of France, 1871–1914* (Cambridge: MIT Press, 1967); Douglas Porch, *March to the Marne: The French Army, 1871–1914* (Cambridge: Cambridge University Press, 1981); Snyder, *Ideology of the Offensive*, pp. 41–106; Arthur J. Marder, *From Dreadnought to Scapa Flow: The Royal Navy in the Fisher Era, 1904–19*, vol. 1: *The Road to War, 1904–1914* (London: Oxford University Press, 1961), pp. 3–11, 28–70; and William C. Fuller, *Civil-Military Conflict in Imperial Russia, 1881–1914* (Princeton: Princeton University Press, 1985).

33. Joseph Schumpeter, *Imperialism and Social Classes*, trans. Heinz Norden (Oxford: Blackwell, 1951), remains the classic treatment of this phenomenon.

34. See Gordon A. Craig, *The Politics of the Prussian Army, 1640–1945* (London: Oxford University Press, 1964), pp. 217–55; Fritz Stern, *The Politics of Cultural Despair: A Study in the Rise of German Ideology* (Berkeley: University of California Press, 1961); and Kenneth D. Barkin, *The Controversy over German Industrialization, 1890–1902* (Chicago: University of Chicago Press, 1970).

35. Fritz Fischer, *War of Illusions* (New York: Norton, 1975), p. 81.

36. See Craig, *Politics of the Prussian Army*, pp. 136–79, on the constitutional struggle, and pp. 217–55, on the army as a state within a state. Otto Pflanze, *Bismarck and the Development of Germany*, vol. 1: *The Period of Unification, 1815–71* (Princeton: Princeton University Press, 1963), remains the best overall treatment of the constitutional crisis from the vantage point of Bismarck.

37. Holger H. Herwig, "Imperial Germany," in May, *Knowing One's Enemies*, pp. 62–97, provides a useful description of army-navy relations and their implications for intelligence and strategy.

38. Klaus Knorr, "Controlling Nuclear War," *International Security* 9 (Spring 1985), pp. 79–98.

39. This perspective is persuasively presented by Timothy J. Colton, *Commissars, Commanders, and Civilian Authority: The Structure of Soviet Military Politics* (Cambridge: Harvard University Press, 1979). The more traditional interpretation depicted civil-military relations as conflict prone but the Party as generally successful in suppressing such conflict. See, for example, Thomas W. Wolfe, "The Military," in Allan Kassof, ed., *Prospects for Soviet Society* (New York: Praeger, 1968), pp. 112–42; Robert Conquest, *Power and Policy in the USSR* (New York: St. Martin's, 1961), p. 330; and Roman Kolkowicz, *The Soviet Military and the Communist Party* (Princeton: Princeton University Press, 1967). The traditional view of U.S. civil-military relations is best expressed by Samuel P. Huntington, *The Soldier and the State* (Cambridge: Harvard University Press, 1957). His model of civilian control, based on the American experience, prescribes keeping the soldiers politically neutral by segregating political and military roles in government and restricting military participation to offering advice about the military implications of policies under consideration. Liberal revisionists challenged this view in the 1960s, arguing, on the basis of their understanding of the Vietnam war, that the military had assumed an increasingly important role in policy making, pushing successive administrations to pursue more interventionist foreign policies. See Arthur Schlesinger, Jr., *The Crisis of Confidence* (Boston: Houghton Mifflin, 1969), and John Kenneth Galbraith, *How to*

Control the Military (New York: New American Library, 1969). A more reasoned argument is found in Richard K. Betts, *Soldiers, Statesmen, and Cold War Crises* (Cambridge: Harvard University Press, 1977), who found the greatest pressure from American soldiers is not on whether to use force but on how to use it.

40. Transfers of weapons from the Atomic Energy Commission to the military continued throughout the Eisenhower administration. By 1961, only 10 percent remained under civilian control. On the evolution of American nuclear weapons policy and targeting in the 1950s see David Alan Rosenberg, "The Origins of Overkill: Nuclear Weapons and American Strategy, 1945–1960," *International Security* 7 (Spring 1983), pp. 3–71; Desmond Ball, *Targeting for Strategic Deterrence*, Adelphi Paper no. 185 (London: Internationl Institute of Strategic Studies, 1983); and Paul Bracken, *The Command and Control of Nuclear Forces* (New Haven: Yale University Press, 1983), pp. 180–82.

41. Gregg Herken, *Counsels of War* (New York: Knopf, 1985), pp. 81–83, 96–98; Rosenberg, "The Origins of Overkill," pp. 37–38; and Thomas Powers, "Nuclear Winter and Nuclear Strategy," *Atlantic*, November 1984, pp. 53–64, who quotes LeMay.

42. George Kistiakowsky, *A Scientist at the White House* (Cambridge: Harvard University Press, 1980), pp. 399–400, 414.

43. See Herken, *Counsels of War*, pp. 143–45; Desmond Ball, *Politics and Force Levels: The Strategic Missile Program of the Kennedy Administration* (Berkeley: University of California Press, 1980), pp. 119, 190–91; and Fred Kaplan, *The Wizards of Armageddon* (New York: Simon & Schuster, 1983), pp. 270–72.

44. See Ball, *Targeting for Strategic Deterrence*, pp. 10–15; Ball, *Politics and Force Levels*, pp. 190–92, 206; and Herken, *Counsels of War*, pp. 156–65.

45. Paul Bracken, "The Political Command and Control of Nuclear Forces" (working paper, Yale University, New Haven, January 1984), pp. 25–26.

46. So argue Betts, *Soldiers, Statesmen, and Cold War Crisis*, pp. 5–12; Leslie H. Gelb with Richard K. Betts, *The Irony of Vietnam: The System Worked* (Washington, D.C.: Brookings, 1979); and Lebow, *Between Peace and War*, pp. 285–87.

47. The reluctance of the services to share information with Agency analysts was due only in part to concern for security. The more important motive was a desire to keep the CIA from conducting "net assessments," that is, overall evaluations of the strategic balance. The Defense Department and the services, especially the air force, insisted that this was their prerogative. One reason they did so was their knowledge that a CIA assessment of the strategic balance would be less extreme than theirs and insufficiently supportive of service budget requests.

48. Quoted in Thomas Powers, "Choosing a Strategy for World War III," *Atlantic*, November 1982, pp. 82–110.

49. Richard H. Ellis, "Building a Plan for Peace: The Joint Strategic Target Planning Staff" (Offut Air Force Base, Neb.: Joint Strategic Planning Staff, 1980), pp. 6–7. Ellis states that formal strategic targeting guidance was not formulated above the JCS level between 1961–62 and 1974.

50. On PD-59 see the 16 September 1980 testimony of Secretary of Defense Harold Brown before U.S. Congress, Senate, *Hearing before the Committee on Foreign Relations of the U.S. Senate on Presidential Directive 59*, 96th Cong., 2d sess. (Washington, D.C., 1981), pp. 6–10, and Walter Slocombe, "The Countervailing Strategy," *International Security* 5 (Spring 1981), pp. 18–27. For developments during the Reagan administration, see Leon Sloss and Marc Dean Millot, "U.S. Nuclear Strategy in Evolution," *Strategic Review* 12 (Winter 1984), pp. 19–28, and Jeffrey Richelson, "PD 59,

NSDD-13, and the Reagan Strategic Modernization Program," *Journal of Strategic Studies* 6 (June 1983), pp. 125–46.

51. Quoted in Daniel Ford, *The Button: The Pentagon's Strategic Command and Control System* (New York: Simon & Schuster, 1985), p. 89.

52. Reported by Powers, "Nuclear Winter and Nuclear Strategy," pp. 63–64, and "What's Worse Than the MX?" *Washington Post*, 31 March 1985, pp. K1–2; Lee Dye, "Top Officials Need Training in War Crisis, Panel Says," *Los Angeles Times*, 31 May 1985, p. 3.

53. Elmo R. Zumwalt, Jr., *On Watch: A Memoir* (New York: Quadrangle, 1976), p. 481.

54. According to Bill Gulley, former director of the White House Military Office, "No new President in my time ever had more than one briefing on the contents of the Football, and that was before each one took office, when it was one briefing among dozens. Not one President . . . ever got an update on the contents of the Football, although material in it is changed constantly." Gulley with Mary Ellen Reese, *Breaking Cover* (New York: Simon & Schuster, 1980), pp. 178–81.

55. This exercise is described by Robert Rosenberg, a former member of the National Security Council staff, in "The Influence of Policy Making on C^3I," in *Seminar on Command, Control, Communications, and Intelligence* (Cambridge: Center for Information Policy Research, Harvard University, 1981), p. 60.

56. Retired general Brent Scowcroft concurs. "Presidents," he says, "are reluctant to face up to this part of their responsibilities in the sense of being willing to participate in exercises, discussions, and so on, about how [the nuclear command and control system works]. It's been my experience that the first couple of days there's a kind of feeling of stark terror when, all of a sudden, they realize, you know, this is all theirs and they're the ones that have to make the decision. But then, as they get into it, it's inherently distasteful to look at, to think about, it's always something that can be done tomorrow rather than today." Quoted in Ford, *The Button*, pp. 90–91. For a discussion of defensive avoidance, see Janis and Mann, *Decision-Making*, pp. 15, 55–56.

57. I elaborate upon this argument in Richard Ned Lebow, "Windows of Opportunity: Do States Jump through Them?" *International Security* 9 (Summer 1984), pp. 147–86.

58. Herken, *Counsels of War*, p. 154.

59. Richard De Lauer reports that national security advisers, like the president, shunned a visit to the National Command Post in the basement of the White House. "I guess President Carter was the first President ever to visit the National Command Post and sit down where he was supposed to sit and at least be briefed on what it all means. President Nixon never did, Johnson never did, and some of the security advisors, like Kissinger, never went down there." Quoted in Ford, *The Button*, p. 89.

60. Lee Dye, "Top Officials Need Training in War Crisis," *Los Angeles Times*, 31 May 1985, p. 3; interview with Joseph Nye, Cambridge, Mass., 17 September 1985.

61. Ford, *The Button*, p. 92, reports an interview with an unidentified senior Reagan official who confided that "the Secretary of State . . . probably knows less about the S.I.O.P. options than the President does" and that Weinberger was also quite ignorant. "Weinberger," according to this official, "looks at his job like a college president. He doesn't see himself as running the intellectual side of the Pentagon. His job is raising money. He's unlike Harold Brown, a physicist, a technician. Weinberger and Reagan together—there's no knowledge to share."

62. Colton, *Commissars, Commanders, and Civilian Authority*, p. 55.

63. Quoted in John Newhouse, *Cold Dawn: The Story of SALT* (New York: Holt, Rinehart & Winston, 1973), p. 192. Raymond L. Garthoff, "The Soviet Military and SALT," in Jiri Valenta and William Potter, eds., *Soviet Decisionmaking for National Security* (London: Allen & Unwin, 1984), pp. 152–53, argues that Newhouse erroneously attributes Ogarkov's concern to discussing Soviet military hardware in front of his civilian colleagues when in fact Ogarkov was objecting to a discussion of Soviet military operational concepts within the context of the SALT negotiations.

64. Noted by Igor Glagolev, "The Soviet Decision-Making Process in Arms Control Negotiations," *Orbis* 27 (Winter 1978), pp. 44–51. Another example of this kind concerns a Soviet scientist, working in a classified military research unit, who requested information on design features of different Soviet aircraft in order to prepare an assigned report. "The First Department [security] of the military unit supplied brief descriptions of the aircraft, each covering several pages. There would be nothing out of the ordinary in this fact if each of the descriptions had not carried a note at the top of the sheet to the effect that it was gleaned from material initially published in the United States, subsequently acquired and translated into Russian. Thus American technical analysis of Soviet aircraft provided a working description of Soviet aircraft for Soviet scientists." Irina Dunskaya, *Security Practices at Soviet Scientific Research Facilities* (Falls Church, Va.: Delphic Associates, 1983), p. 139.

65. On this subject see also Edward L. Warner III, *The Military in Contemporary Soviet Politics: An Institutional Analysis* (New York: Praeger, 1977), p. 25; Harriet Fast Scott and William F. Scott, *The Armed Forces of the USSR* (Boulder, Colo.: Westview, 1979), pp. 108–9; Arthur J. Alexander, *Decision-Making in Soviet Weapons Procurement*, Adelphi Papers nos. 147 and 148 (London: International Institute of Strategic Studies, 1978), p. 40; and David Holloway, *The Soviet Union and the Arms Race* (New Haven: Yale University Press, 1983), p. 110.

66. Quoted in Mohamed H. Heikal, *The Sphinx and the Commissar* (New York: Harper & Row, 1978), p. 129.

67. Cf. William Moffitt and Ross Stagner, "Perceptual Rigidity and Closure as a Function of Anxiety," *Journal of Abnormal and Social Psychology* 52 (May 1956), pp. 354–57; Richard S. Lazarus and Susan Folkman, *Stress, Appraisal, and Coping* (New York: Springer, 1984); and Ole R. Holsti, "Cognitive Dynamics and Images of the Enemy," in David Finlay, Holsti, and Richard Fagen, eds., *Enemies in Politics* (Chicago: Rand-McNally, 1967), pp. 25–96.

68. Corelli Barnett, *The Swordbearers: Supreme Command in the First World War* (Bloomington: Indiana University Press, 1963), p. 36.

69. Helmuth von Moltke, *Erinnerungen, Briefe, Dokumente, 1877–1916*, ed. Eliza von Moltke (Stuttgart: Kommende Tag, 1922), pp. 19–20. Lebow, *Between Peace and War*, pp. 232–42, makes the case for the psychological roots of Moltke's rigidity. Levy, "Organizational Routines and the Causes of War," reviews the evidence for this argument and the organizational and political explanations for Moltke's behavior.

70. Yuri Danilov, *La Russie dans la Guerre Mondiale, 1914–1917*, trans. Alexandre Kaznakov (Paris: Payot, 1927), pp. 33, 292–93; W. A. Sukhomlinov, *Erinnerungen* (Berlin: Reimar Hobbing, 1924), pp. 364–65; von Schilling, *How the War Began*, p. 117; Snyder, *Ideology of the Offensive*, pp. 165–98.

71. Albertini, *Origins of the War of 1914*, 3: 253.

72. Quoted in Kaplan, *Wizards of Armageddon*, p. 298.

73. U.S. Congress, House, Subcommittee of the Committee on Appropriations,

Hearings on Department of Defense Appropriations for 1961, 86th Cong., 2d sess. (Washington, D.C., 1960), pt. 7, p. 88.

74. Kaplan, *Wizards of Armageddon*, pp. 370–71. Also see John Edwards, *Super Weapon: The Making of the MX* (New York: Norton, 1982), pp. 67–68.

75. Other nuclear powers may confront the same problem. Denis Healy reported that, to his knowledge, he was the first British minister of defense to ask to see the strategic war plans. What he found appalled him: none of the options would have been politically acceptable to the government. Columbia University seminar, 14 October 1971.

76. Stephen M. Meyer, "Soviet Perspectives on the Paths to Nuclear War," in Graham T. Allison, Albert Carnesale, and Joseph S. Nye, Jr., *Hawks, Doves, and Owls: An Agenda for Avoiding Nuclear War* (New York: Norton, 1985), pp. 167–205.

77. See Richard Ned Lebow, "The Soviet Offensive in Europe: The Schlieffen Plan Revisited?" *International Security* 9 (Spring 1985), pp. 44–78, for a description and critique of this strategy.

78. Jack Snyder, "Civil-Military Relations and the Cult of the Offensive in 1914 and 1984," *International Security* 9 (Summer 1984), pp. 108–46.

79. Benjamin S. Lambeth, *Selective Nuclear Options in American and Soviet Strategic Policy* (Santa Monica, Calif.: Rand, 1976), p. vi.

80. See, for example, Dennis M. Gormley and Douglas M. Hart, "Soviet Views on Escalation," *Washington Quarterly* 7 (Fall 1984), pp. 9–17.

81. See Ball, *Targeting for Strategic Deterrence*, pp. 8–15.

82. In interviews, officials from the Kennedy through the Reagan administrations have with striking frequency complained that in the absence of existing files, they have had to start their work from scratch. They have also stated their regret that upon assuming office, they failed to solicit the advice of their predecessors.

83. See Janis and Mann, *Decision-Making*, pp. 15, 55–56.

84. Ibid., pp. 74–95.

85. Ibid., pp. 76–79.

86. I have documented this in Lebow, *Between Peace and War*, pp. 119–47.

87. Norman Stone, *The Eastern Front, 1914–1917* (New York: Scribners, 1975), pp. 70–92, offers a good description of Austria's military dilemma and the unrealistic ways in which the general staff sought to deal with it.

88. This problem is treated in Norman Stone, "Moltke-Conrad: Relations between the Austro-Hungarian and German General Staffs, 1900–1914," *Historical Journal* 9, 2 (1966), pp. 201–28. The Germans had a reciprocal illusion. Once they became committed to an offensive against France with only a small covering force to defend themselves in the east, they came to rely upon an Austrian offensive against Russia. A review of the relevant documents would reveal, I suspect, that for this reason they consistently exaggerated the extent of the effort that Austria was prepared to make against Russia.

89. Snyder, *Ideology of the Offensive*, pp. 157–98, analyzes this dilemma in detail and demonstrates how motivated bias led to a Russian overcommitment to the offensive.

90. For differing assessments of the internal situation, see Leopold Haimson, "The Problem of Social Stability in Urban Russia, 1905–1917," *Slavic Review* 23 (December 1964), pp. 619–42, and 24 (March 1965), pp. 1–22; Hans Rogger, "Russia in 1914," in Walter Laqueur and George L. Mosse, eds., *1914: The Coming of the First World War* (New York: Harper & Row, 1966), pp. 229–53; and Arno J. Mayer,

"Domestic Causes of the First World War," in Leonard Krieger and Fritz Stern, eds., *The Responsibility of Power: Historical Essays in Honor of Hajo Holborn* (Garden City, N.Y.: Doubleday, 1969), pp. 308–24.

91. See Lebow, *Between Peace and War*, and "Miscalculation in the South Atlantic: The Origins of the Falklands War," in Jervis, Lebow, and Stein, *Psychology and Deterrence*, pp. 89–124; Stein, "Calculation, Miscalculation, and Conventional Deterrence I: The View From Cairo," and "II: The View from Jerusalem," in the same volume, pp. 34–59, 60–88.

92. These are described in Richard Ned Lebow, "The Cuban Missile Crisis: Reading the Lessons Correctly," *Political Science Quarterly* 98 (Fall 1983), pp. 431–58.

93. See Elie Abel, *The Missile Crisis* (Philadelphia: Lippincott, 1966), pp. 9–10, and Robert F. Kennedy, *Thirteen Days: A Memoir of the Cuban Crisis* (New York: Norton, 1969), pp. 24–26.

94. Allison, *Essence of Decision*, p. 42. According to Denis Healy, former British defense minister, Kennedy's handling of the crisis, including the events leading up to it, "could be cited as a model in any textbook of diplomacy." Quoted in Henry H. Pachter, *Collision Course: The Cuban Missile Crisis and Coexistence* (New York: Praeger, 1963), p. 87. Later in his book, Allison, pp. 232–37, when employing the bureaucratic perspective to analyze the Soviet decision to put missiles into Cuba, backs away from his earlier judgment.

95. See Arnold L. Horelick and Myron Rush, *Strategy, Power and Soviet Foreign Policy* (Chicago: University of Chicago Press, 1966), p. 141; Roger Hilsman, *To Move a Nation: The Politics of Foreign Policy in the Administration of John F. Kennedy* (Garden City, N.Y.: Doubleday, 1967), pp. 200–202; Michel Tatu, *Power in the Kremlin: From Khrushchev's Decline to Collective Leadership*, trans. Helen Katel (London: Collins, 1969), p. 231; Ronald Steel, "End Game," *New York Review of Books*, 13 March 1969, p. 229; Allison, *Essence of Decision*, pp. 52–56; and George and Smoke, *Deterrence in American Foreign Policy*, pp. 461–62.

96. Allison, *Essence of Decision*, pp. 237–44.

97. For the most convincing alternative interpretations, see Allison, *Essence of Decision*, pp. 235–37, and George and Smoke, *Deterrence in American Foreign Policy*, pp. 464–66, 469–70, 488–89. The latter stresses the Soviet approach to risk calculation and risk acceptance.

98. See Hilsman, *To Move a Nation*, p. 184; U.S. Congress, House, Subcommittee of the Committee on Appropriations, *Hearings on Department of Defense Appropriations*, 86th Cong., 1st sess. (Washington, D.C., 1963), p. 25; and Allison, *Essence of Decision*, pp. 235–37.

99. See Kennedy, *Thirteen Days*, pp. 31, 34, 38–39; Arthur Schlesinger, Jr., *A Thousand Days: John F. Kennedy in the White House* (Boston: Houghton Mifflin, 1965), pp. 806–7; Theodore C. Sorensen, *Kennedy* (New York: Harper & Row, 1965), p. 684; and Allison, *Essence of Decision*, pp. 59–61, 203–4.

100. Quoted in *New York Times*, 11 June 1963; Kennedy, *Thirteen Days*, p. 126.

101. Herken, *Counsels of War*, p. 168, claims that this was Curtis LeMay.

102. Sloan Foundation interviews, quoted in Herken, *Counsels of War*, p. 168.

103. Dean Acheson, "Homage to Plain Dumb Luck," *Esquire*, February 1969, pp. 44–46, 76–77.

104. Quoted in Edward Weintal and Charles Bartlett, *Facing the Brink: An Intimate Study of Crisis Diplomacy* (New York: Scribners, 1967), pp. 54–55.

105. For example, Herman Kahn, *On Escalation* (New York: Praeger, 1965), pp.

74–82; Schelling, *Arms and Influence*, pp. 80–83; and Albert and Roberta Wohlstetter, *Controlling the Risks in Cuba*, Adelphi Paper no. 17 (London: International Institute of Strategic Studies, 1965), p. 16.

106. Wohlstetter and Wohlstetter, *Controlling the Risks in Cuba*, p. 16.

107. For a fuller critique of this way of thinking, see Richard Ned Lebow, "Misconceptions in American Strategic Assessment," *Political Science Quarterly* 97 (Summer 1982), pp. 187–206, and "Conclusions," in Jervis, Lebow, and Stein, *Psychology and Deterrence*, pp. 203–32.

108. Quoted in Sorensen, *Kennedy*, p. 677.

109. On this point see Jerome H. Kahan and Anne K. Long, "The Cuban Missile Crisis: A Study of Its Strategic Context," *Political Science Quarterly* 87 (December 1972), pp. 564–90; and Allison, *Essence of Decision*, pp. 56–61.

110. See, for example, Richard Pipes, "Why the Soviet Union Thinks It Could Fight and Win a Nuclear War," *Commentary* 64 (July 1977), pp. 21–34; Edward N. Luttwak, "After Afghanistan, What?" *Commentary* 69 (April 1980), pp. 40–49; and William R. Graham and Paul H. Nitze, "Variable U.S. Strategic Missile Forces for the Early 1980s," in William R. Van Cleave and W. Scott Thompson, *Strategic Options for the Early Eighties: What Can Be Done?* (White Plains, Md.: Automated Graphic Systems, 1979).

111. It could also make American leaders more cautious than they might otherwise be, because of their perception that the military balance has swung so much in the direction of the Soviets since the 1960s.

112. A good outline is Richard Nisbett and Lee Ross, *Human Inference: Strategies and Shortcomings of Social Judgment* (Englewood Cliffs, N.J.: Prentice-Hall, 1980), pp. 120–38.

113. Ibid., pp. 273–95.

114. The 1973 crisis, arising out of the Middle East war, also involved a nuclear alert, but most authorities consider that in comparison to 1962 it was not a serious crisis.

115. Research indicates that up to a certain point, stress can have a beneficial influence upon decision making. These findings are discussed by Joseph de Rivera, *The Psychological Dimension of Foreign Policy* (Columbus, Ohio: Bobbs-Merrill, 1968), pp. 150–51; Dean G. Pruitt, "Definition of the Situation as a Determinant of International Action," in Herbert C. Kelman, ed., *International Behavior* (New York: Holt, Rinehart & Winston, 1965), pp. 391–432; Ole R. Holsti and Alexander L. George, "The Effects of Stress on the Performance of Foreign Policy-Makers," in C. P. Cotter, ed., *Political Science Annual: An International Review* (Indianapolis: Bobbs-Merrill, 1975), pp. 255–319; and Janis and Mann, *Decision-Making*.

116. For a discussion of some of this literature, see Irving L. Janis, *Psychological Stress* (New York: Wiley, 1958); Holsti and George, "Effects of Stress"; Janis and Mann, *Decision-Making;* and Thomas C. Wiegele, "Decision-Making in an International Crisis: Some Biological Factors," *International Studies Quarterly* 17 (September 1973), pp. 295–335.

117. See Ole R. Holsti, *Crisis, Escalation, War* (Montreal: McGill-Queens University Press, 1972); Irving L. Janis, *Groupthink* (Boston: Houghton Mifflin, 1982); and Lebow, *Between Peace and War*, chaps. 4 and 6.

118. Alexander L. George, "The Impact of Crisis-induced Stress on Decisionmaking," paper prepared for Institute of Medicine Symposium on the Medical Aspects of Nuclear War, National Academy of Sciences, Washington, D.C., 20–22 September 1985. Also see his earlier "Adaptation to Stress in Political Decision Making:

The Individual, Small Group and Organizational Contexts," in G. V. Coelho, David A. Hamburg, and J. E. Adams, eds., *Coping and Adaptation* (New York: Basic, 1974), pp. 176–224.

119. According to the American Psychiatric Association's *Diagnostic and Statistical Manual of Mental Disorders*, 3d ed. (Washington, D.C., 1980), p. 253, the essential feature of a dissociative disorder "is a sudden, temporary alteration in the normally integrative functions of consciousness, identity, or motor behavior. If the alteration occurs in consciousness, important personal events cannot be recalled. If it occurs in identity, either the individual's customary identity is temporarily forgotten and a new identity is assumed, or the customary feeling of one's own reality is lost and replaced by a feeling of unreality. If the alteration occurs in motor behavior, there is also a concurrent disturbance in consciousness or identity, as in the wandering that occurs during a Psychogenic Fugue."

120. The breakdowns of Kaiser Wilhelm, Stalin, Nehru, and Nasser are documented in Lebow, *Between Peace and War*, pp. 135–45, 283–85. On Eden see Hugh Thomas, *Suez* (New York: Harper & Row, 1966), pp. 149, 162. For Rabin, see Ezer Weizman, "Aide-Memoire," *Jerusalem Post*, 23 April 1974; Yitzhak Rabin, *The Rabin Memoirs* (Boston: Little, Brown, 1979), pp. 79–82; Michael Brecher with Benjamin Geist, *Crisis and Decision: Israel, 1967, 1973* (Berkeley: University of California Press, 1980), pp. 345–47; and Janice Gross Stein and Raymond Tanter, *Rational Decision-Making: Israel's Security Choices, 1967* (Columbus: Ohio State University Press, 1980), pp. 159, 172, 176.

121. Anwar L. Sadat, *In Search of Identity: An Autobiography* (New York: Harper & Row, 1977), p. 180.

122. See Lebow, *Between Peace and War*, pp. 119–47, for a case study.

123. See N. G. Kuznetsov, "At Naval Headquarters," and I. V. Tuilenev, "At Moscow District Headquarters," in Seweryn Bialer, ed., *Stalin and His Generals* (New York: Pegasus, 1969), pp. 190–200, 200–203; A. M. Nekrich, "June 22, 1941," in Vladimir Petrov, ed., *June 22, 1941: Soviet Historians and the German Invasion* (Columbia: University of South Carolina Press, 1968), pp. 240–45; and Nikita S. Khrushchev, *Khrushchev Remembers*, ed. Edward Crankshaw, 2 vols. (Boston: Little, Brown, 1970), 1: 126–35. The most recent primary source if Valentin Berezhkov, *History in the Making: Memoirs of World War II Diplomacy*, trans. Dudley Hagen and Barry Jones (Moscow: Progress, 1983), pp. 144–50, who also testifies to Stalin's unwillingness to contemplate a German invasion, describes some of his efforts to hinder military preparedness on the eve of the invasion, and speculates about the psychological reasons for his need to behave in this way. Berezhkov also describes a meeting with Stalin in the winter of 1941 and his own shock at the extent to which the Soviet leader seemed physically and psychologically affected by the ordeal of the war. For Western secondary sources, see Ammon Sella, "Barbarossa: Surprise Attack and Communication," *Journal of Contemporary History* 13 (July 1978), pp. 555–84, and Albert Seaton, *Stalin as a Military Commander* (New York: Praeger, 1976), pp. 95–101.

124. Ivan Maisky in *Novyi Mir*, December 1964, p. 163, quoted in Adam Ulam, *Expansion and Coexistence: The History of Soviet Foreign Policy, 1917–1967* (New York: Praeger, 1968), p. 315. See also Khrushchev, *Khrushchev Remembers*, 1: 166, and Harrison Salisbury, *The 900 Days: The Seige of Leningrad* (New York: Harper & Row, 1969), pp. 57–81.

125. Georgii Zhukov, *The Memoirs of Marshal Zhukov* (New York: Delacorte, 1971), pp. 234–36, 238.

126. Kuznetsov, "At Naval Headquarters," p. 194.

127. Cf. Vojtech Mastny, "Stalin and the Prospects of a Separate Peace in World War II," *American Historical Review* 77 (December 1972), pp. 1365–88. Stalin's initial effort to treat the German invasion as a limited war with limited aims is also commented on by Franz Halder, *The Private War Journal of Generaloberst Franz Halder*, 9 vols. (Nuremberg: Office of Chief of Counsel for War Crimes, 1950), 6: 162.

128. This is reported by George, "Impact of Crisis-Induced Stress," p. 16.

129. This was developed by Harvey D. Reed and Irving L. Janis, "Effects of a New Type of Psychological Treatment on Smokers' Resistance to Warnings about Health Hazards," *Journal of Consulting and Clinical Psychology* 42 (October 1974), p. 748; also see Janis and Leon Mann, "Coping With Decisional Conflict," *American Scientist* 64 (November–December 1976), pp. 657–66. The awareness of rationalization technique has much in common with the insight technique developed by Daniel Katz, Irving Sarnoff, and Charles G. McClintock, "Ego Defense and Attitude Change," *Human Relations* 9 (February 1956), pp. 27–46, and Milton Rokeach's awareness of inconsistency between values and actions, "Long-Range Experimental Modification of Values, Attitudes, and Behavior," *American Psychologist* 26 (May 1971), pp. 453–59.

130. See Herbert C. Kelman, "Attitude Change as a Function of Response Restriction," *Human Relations* 6 (August 1953), pp. 185–214; Bert T. King and Irving L. Janis, "The Influence of Role Playing on Opinion-Change," *Journal of Abnormal and Social Psychology* 49 (April 1954), pp. 211–18, and "Comparison of the Effectiveness of Improvised vs. Non-Improvised Role Playing in Producing Opinion Changes," *Human Relations* 9 (Ausust 1956), pp. 177–86; Phillip G. Zimbardo, "The Effect of Effort and Improvisation on Self-Persuasion Produced by Role-Playing," *Journal of Experimental Social Psychology* 1 (May 1965), pp. 103–20; Janis and Mann, *Decision-Making*, pp. 348–65, 403, 441–42; and Michael Toomey, "Conflict Theory Approach to Decision-Making Applied to Alcoholics," *Journal of Personality and Social Psychology* 24 (November 1972), pp. 199–206. The Japanese had a contingency plan for calling off their attack on Pearl Harbor had their carrier force been discovered by the Americans. It is conceivable that Khrushchev had also left himself a line of retreat had Kennedy mounted a more timely or more private response than he did. Evidence to this effect—although hardly likely to surface—would require reconsideration of the defensive avoidance hypothesis.

131. On combat see Irving L. Janis, "Problems Related to the Control of Fear in Combat," in S. Stouffer et al., *The American Soldier*, vol. 2: *Combat and Its Aftermath* (Princeton: Princeton University Press, 1949), pp. 192–214. S. J. Rachman, *Fear and Courage* (San Francisco: Freeman, 1978), offers a critical review of the literature on this subject. On stress innoculation in general see Janis, *Air War and Emotional Stress* (New York: McGraw-Hill, 1951), and *Psychological Stress* (New York: Wiley, 1958), and Donald Meichenbaum, *Cognitive-Behavior Modification: An Integrative Approach* (New York: Plenum, 1977), who coined the term. For the use of the technique in surgery, see L. Egbert, G. Battit, C. Welch, and M. Bartlett, "Reduction of Postoperative Pain by Encouragement and Instruction of Patients," *New England Journal of Medicine* 270 (April 1964), pp. 825–27; J. E. Johnson, "The Influence of Purposeful Nurse-Patient Interaction on the Patient's Postoperative Course," in *Exploring Medical and Surgical Nursing Practice*, American Nursing Association Monograph Series no. 2 (New York: American Nursing Association, 1966); and David T. A. Vernon and Douglas A. Bigelow, "The Effect of Information about a Potentially Stressful Situation on Responses to Stress Impact," *Journal of Personality and Social Psychology* 29 (January 1974), pp. 50–59. On the technique and childbirth, see F. Lamaze,

Painless Childbirth: Psychoprophylactic Method (London: Burke, 1958), and G. Dick-Read, *Childbirth without Fear: The Principles and Practices of Natural Childbirth*, 2d rev. ed. (New York: Harper & Row, 1959). A critical review of this literature is provided by Janis, "Stress Innoculation in Health Care: Theory and Research," in Donald Meichenbaum and Matt Jaremko, eds., *Stress Reduction and Prevention* (New York: Plenum, 1983), pp. 67–99.

132. See Irving L. Janis and Leon Mann, "Effectiveness of Emotional Role-Playing in Modifying Smoking Habits and Attitudes," *Journal of Experimental Research in Personality* 1 (October 1965), pp. 84–90; Toomey, "Conflict Theory Approach to Decision-Making Applied to Alcoholics"; Mann and Janis, "A Follow-Up Study on the Long-Term Effects of Emotional Role Playing," *Journal of Personality and Social Psychology* 8 (April 1968), pp. 339–42; and Gerald L. Clore and Katherine L. McMillan, "Emotional Role Playing, Attitude Change, and Attraction toward a Disabled Person," *Journal of Personality and Social Psychology* 23 (July 1972), pp. 105–11.

133. See Irving L. Janis and Donna LaFlamme, "Effects of Outcome Psychodrama as a Supplementary Technique in Marital and Career Counseling," in Janis, ed., *Counseling on Personal Decisions: Theory and Research on Short-Term Helping Relationships* (New Haven: Yale University Press, 1982), pp. 305–12. The authors found role playing to have a negative effect on stress reduction.

134. Evidence for this also comes from attempts at "brainwashing." During the Korean War the Chinese used role playing without much success in an attempt to change the political attitudes of American prisoners of war. Former prisoners report that while verbalizing good procommunist arguments, they inwardly thought of counterarguments and derisive epithets for their Chinese group leaders. See E. H. Schein, "The Chinese Indoctrination Program for Prisoners of War: A Study of Attempted 'Brainwashing,'" *Psychiatry* 19 (May 1956), pp. 149–72, and Robert J. Lifton, *Thought Reform and the Psychology of Totalism: A Study of "Brainwashing" in China* (New York: Norton, 1961).

135. Janis, *Counseling on Personal Decisions*, chap. 20, reviews research findings on specific personality variables and responsiveness to stress innoculation.

136. Other cases are discussed in Lebow, *Between Peace and War*, and Jervis, Lebow, and Stein, *Psychology and Deterrence*.

5. *Toward Crisis Stability*

1. Richard Ned Lebow, *Between Peace and War: The Nature of International Crisis* (Baltimore: Johns Hopkins University Press, 1981), pp. 254–63, reviews the evidence in support of this growing sense of fatalism.

2. Wolfgang Mommsen, "The Debate on German War Aims," *Journal of Contemporary History* 1 (July 1966), pp. 47–74.

3. Count Alfred von Schlieffen, "Der Krieg in der Gegenwart," *Deutsche Revue* (1909), pp. 13–24, quoted in Jonathan Steinberg, "The Copenhagen Complex," in Walter Laqueur and George L. Mosse, eds., *1914: The Coming of the First World War* (New York: Harper & Row, 1966), p. 38.

4. Fritz Fischer, *War of Illusions* (New York: Norton, 1975), p. 402, quoting from Jagow's papers.

5. Lebow, *Between Peace and War*, pp. 256–59, and "Windows of Opportunity: Do States Jump through Them?" *International Security* 9 (Summer 1984), pp. 147–86,

review the evidence in this connection and discuss some of the alternative explanations.

6. Walter Goerlitz, ed., *Der Kaiser . . . Aufzeichnungen des Chefs des Marinekabinetts Admiral Georg Alexander v. Müller, 1914–1918* (Göttingen: Munsterschmidtverlag, 1959), p. 124.

7. Immanuel Geiss, "The Crisis of July 1914," in Laqueur and Mosse, *1914*, pp. 78–79, citing documents in Jagow's papers; and Volker R. Berghahn, *Germany and the Approach of War in 1914* (New York: St. Martin's, 1973), pp. 187–88.

8. See Luigi Albertini, *The Origins of the War of 1914*, trans. Isabella M. Massey, 3 vols. (Oxford: Oxford University Press, 1952), 2: 496–97, 673–77, and 3: 6–14; Gerhard Ritter, *Sword and Scepter: The Problem of Militarism in Germany*, 4 vols., trans. Heinz Norden (Coral Gables: University of Miami Press, 1970), 2: 239–63.

9. I discussed this question in Chapter 4 and reviewed the relevant literature in Richard Ned Lebow, "The Cuban Missile Crisis: Reading the Lessons Correctly," *Political Science Quarterly* 98 (Fall 1983), pp. 431–58.

10. So argues William Taubman, *Stalin's American Policy: From Entente to Détente to Cold War* (New York: Norton, 1982), p. 72.

11. Cf. *Khrushchev Remembers*, trans. Strobe Talbot (Boston: Little, Brown, 1970), p. 392.

12. On Bethmann-Hollweg's suspicions see Konrad H. Jarausch, *The Enigmatic Chancellor: Bethmann-Hollweg and the Hubris of Imperial Germany* (New Haven: Yale University Press, 1973), p. 58. Lebow, *Between Peace and War*, pp. 247–54, examines pre-1914 German attitudes toward war.

13. See Richard Ned Lebow, "Away from the Abyss: Managing Soviet-American Strategic Rivalry" (forthcoming), chap. 6, for a discussion of Soviet pronouncements on nuclear war and the Western debate about their significance.

14. See ibid., chap. 9, which compares and contrasts the underlying conditions responsible for World War I with today's international environment.

15. The outstanding example of this approach is Bruce G. Blair, *Strategic Command and Control: Redefining the Nuclear Threat* (Washington, D.C.: Brookings, 1985). Blair's objective is to resolve entirely the tension between the need to guarantee retaliation and the need to prevent an accidental or unauthorized launch. His solution is to improve the survivability of command and control, and that of the strategic forces as well, to the point where strict negative control could be maintained well into the period after an enemy attack without jeopardizing the American capability to retaliate. As Blair himself admits, this solution would require "radical adjustments" to the U.S. nuclear posture. In my opinion it is a worthwhile but illusory goal.

16. Reported in Coral Bell, *The Conventions of Crisis: A Study in Diplomatic Management* (New York: Oxford University Press, 1971), p. 2.

17. The pioneering work of Alexander George and his collaborators on this subject can serve as a sturdy foundation on which the rest of us can aspire to build. See especially George, *Managing U.S.-Soviet Rivalry: Problems of Crisis Prevention* (Boulder, Colo.: Westview, 1983).

18. Herman Kahn, *On Thermonuclear War*, 2d ed. (Princeton: Princeton University Press, 1961), pp. 198–99. Richard K. Betts, "Surprise Attack and Preemption," in Graham T. Allison, Albert Carnesale, and Joseph S. Nye, Jr., eds., *Hawks, Doves, and Owls: An Agenda for Avoiding Nuclear War* (New York: Norton, 1985), pp. 54–79, describes an American variant. Lyndon Johnson, he reports, is alleged to have said: "Some people wonder what would happen if I just woke up on the wrong side of

the bed one day, decided I'd had it with the Russians, called the commander of SAC, and said 'General, go get 'em!' You know what the general would say? 'Screw you, Mr. President.'"

19. See, for example, William Prochnau, *Trinity's Child* (Philadelphia: Putnam, 1983).

20. Blair, *Strategic Command and Control*, p. 289, offers this as a goal toward which command and control improvements ought to aim.

21. The forthcoming report of the Cornell University Peace Studies Program–American Academy of Arts and Sciences research project on crisis stability, *Command in Crisis* (New York: Oxford University Press), by Kurt Gottfried and Bruce G. Blair will contain numerous recommendations of this kind.

22. See Bruce Blair, "Solving the Command and Control Problem," *Arms Control Today* 15 (January 1985), pp. 1–9.

23. This is the conclusion reached by Daniel Ford, *The Button: The Pentagon's Strategic Command and Control System* (New York: Simon & Schuster, 1985), pp. 233–42.

24. On the U.S. commitment to ride out a first strike see the 16 September 1980 testimony of Secretary of State Edmund Muskie and Secretary of Defense Harold Brown in U.S. Congress, Senate, *Hearing before the Committee on Foreign Relations of the U.S. Senate on Presidential Directive 59*, 96th Cong. 2d. sess. (Washington, D.C., 1981), pp. 3–5, 6–10; also U.S. Arms Control and Disarmament Agency, *Arms Control Impact Statement for Fiscal Year 1981*, 96th Cong., 2d sess. (Washington, D.C., 1980), p. 19. More recently this commitment was taken as the baseline for strategic analysis by the Scowcroft Commission, *Report of the President's Commission on Strategic Forces* (Washington, D.C.: Department of Defense, April 1983), p. 8.

25. See, for example, Richard L. Garwin, "Launch under Attack to Redress Minuteman Vulnerability," *International Security* 4 (Winter 1979–80), pp. 117–39.

26. For a compelling technical critique of LOW and LUA, see John Steinbruner, "Launch under Attack," *Scientific American* 250 (January 1984), pp. 37–47. LUA is also dismissed as impractical and dangerous by Albert Carnesale and Charles Glaser, "ICBM Vulnerability: The Cures Are Worse Than the Disease," *International Security* 7 (Summer 1982), pp. 70–85.

27. Cf. Peter Pringle and William Arkin, *SIOP: The Secret U.S. Plan for Nuclear War* (New York: Norton, 1983), pp. 112–15, and John Prados, *The Soviet Estimate: U.S. Intelligence Analysis and Soviet Military Strength* (New York: Dial, 1982), pp. 106–10.

28. See testimony of Lt.-Gen. Kelly H. Burke, in U.S. Congress, House, Hearings before a Subcommittee of the House Committee on Appropriations, *Department of Defense Appropriations for 1981*, 96th Cong., 2d sess. (Washington, D.C., 1980), pt. III, p. 1045. See also Blair, *Strategic Command and Control*, pp. 234–38, for a discussion.

29. U.S. Congress, Senate, Appropriations Committee, *Hearings on a Resolution to Approve Funding for the MX Missile*, 98th Cong., 1st sess. (Washington, D.C., 1983), pp. 161–265; Richard Halloran in the *New York Times*, 6 May 1983, pp. A1 and B6. In April 1983 an air force lieutenant colonel, later rebuked by his Pentagon superiors, told a Wyoming chamber of commerce that the United States was moving toward a LOW posture. Fred Iklé, under secretary of defense for policy, issued the following statement in response to the disclosure: "It is our policy not to explain in detail how we would respond to a missile attack, to increase the uncertainties in the minds of Soviet planners. However, the United States does not rely on its capability for launch on warning or launch under attack to ensure the credibility of its deterrent."

Quoted in Hedrick Smith, "Colonel Stirs Questions on M-X Firing Doctrine," *New York Times*, 8 April 1983, p. D15. More recently Secretary of Defense Weinberger denies that the United States had a nuclear "hair trigger," in a letter to Senator William Proxmire, quoted in *Chicago Tribune*, 3 September 1985, p. 4.

30. Michael Olshausen, "On Recalling the Unrecallable: The Militarily Secure, Discretionary Disabling of Nuclear Warheads in Mid-Flight" (unpublished paper, Washington, D.C., 1985), has proposed how this might be done.

31. See Richard K. Betts, *Surprise Attack: Lessons for Defense Planning* (Washington, D.C.: Brookings, 1982), p. 248, and Blair, *Strategic Command and Control*, pp. 235–36.

32. Steinbruner, "Launch under Attack"; Betts, *Surprise Attack*, p. 248, also disparages LOW but argues the feasibility of LUA.

33. The need for any LOW system to rely on fully automated systems to detect and make a decision to respond to attack is recognized by the Department of Defense Advance Research Projects Agency's strategic computing plan. Begun in October 1983, the plan aims to develop a new generation of computing technology for the command and control of strategic forces. Defense Advanced Research Projects Agency, "New Generation Computing Technology: A Strategic Plan for Its Development and Application to Critical Problems in Defense" (Washington, D.C., 28 October 1983), provides an overview of this project. For a critique of the plan and a discussion of the implications of automated decision making, see Severo M. Ornstein, Brian C. Smith, and Lucy A. Suchman, "Strategic Computing," *Bulletin of the Atomic Scientists* 40 (December 1984), pp. 11–15. Herb Lin, "The Development of Software for Ballistic Missile Defense," *Scientific American* 253 (December 1985), pp. 46–53, offers a compelling technical critique of the feasibility of developing reliable software programs for BMD.

34. Department of Defense, *Annual Report for Fiscal Year 1982* (Washington, D.C., 1981), p. 55.

35. Each leg of the triad is assigned a percentage of counterforce and countervalue targets. Under day-to-day conditions of readiness, most of the ICBMs are targeted against Soviet silos and command and control, the bombers are divided between counterforce and countervalue targets, and the submarines are given primarily to countervalue targets. Retaliation in these circumstances would force the NCA to make targeting trade-offs. With the targeting distribution described above, the United States might have very little ability to destroy important countervalue targets if many of the bombers, as is expected, were destroyed at their bases and the NCA had difficulty in communicating with the submarines. To use some of the ICBMs to cover this target set, however, would reduce the number available to strike at counterforce targets. Full generation obviates these kind of trade-offs and permits choice among a wider range of SIOP options with greater confidence in their successful execution. See Desmond Ball, *Targeting for Strategic Deterrence*, Adelphi Paper no. 185 (London: International Institute of Strategic Studies, 1983), especially pp. 17–26, for details.

36. This point was made by Harold Brown, a critic on technical grounds of LOW and LUA. With regard to LUA, he told the Congress: "I think that launch under attack is something that is important to have as an option. . . . Let me put it this way: The Soviets should not be able to count, and I think aren't able to count, on our not doing it, but we surely should not count on being able to. That is uncomfortable, but that is the way it is, and I think that contributes to deterrence." *Hearings on PD 59*, p. 18.

37. I first broached this notion in "Practical Ways to Avoid Superpower Crises," *Bulletin of the Atomic Scientists* 48 (January 1985), pp. 22–28.

38. It is important to distinguish this proposal from the more elaborate crisis management centers proposed by Richard K. Betts and Senator Sam Nunn. I believe that such centers, while they would be helpful, are also impractical. Neither superpower leadership would be prepared to share or alienate any of its authority for managing foreign policy crises. Crisis management centers, regardless of how they were set up and staffed, would be perceived as threats or at least impediments to the freedom of political leaders to manage crises as they saw fit.

39. Douglas M. Hart, "Soviet Approaches to Crisis Management," *Survival* 26 (September–October 1984), pp. 214–22, points out that Soviet conceptions of crisis stability are quite different. He cites Gen. R. Petrov's assertion in *Pravda* (Bratislava), 22 January 1983, p. 4, trans. in FBIS, *Daily Report: Soviet Union,* 26 January 1983, p. A-8, that destabilizing forces are those that "do not have sufficiently reliable communications with headquarters—this fact enhances the probability of their being used without approval—and which require special operational measures which can aggravate tension (for instance the take-off of heavy bombers)." Perhaps it is not entirely incidental that the American definition of crisis stability emphasizes the danger of large and accurate land-based missiles, the Soviet's long suit. Petrov's definition—and he makes this explicit—shifts the onus to SSBNs, long-range bombers equipped with stealth technology, cruise missiles, and forward-deployed Pershing IIs. It is thus equally self-serving.

40. Blair, *Strategic Command and Control,* p. 300.

41. See "A Nuclear Risk Reduction System," excerpts from the "Report of the Nunn/Warner Working Group on Nuclear Risk Reduction" (November 1983), in *Survival* 26 (May–June 1984), pp. 133–36; Richard K. Betts, "A Joint Nuclear Risk Control Center" (Washington, D.C.: Brookings, 1983); William L. Ury, *Beyond the Hotline* (Boston: Houghton Mifflin, 1985); and Paul Bracken, "Accidental War," and Albert Carnesale, Joseph S. Nye, Jr., and Graham T. Allison, "An Agenda for Action," both in Allison, Carnesale, and Nye, *Hawks, Doves, and Owls,* pp. 25–53, 223–46.

42. For an elaboration of this argument, see Lebow, "Practical Ways to Avoid Superpower Crises."

43. According to Ball, *Targeting for Strategic Deterrence,* pp. 17–24.

44. Meyer, "Soviet Perspectives on Paths to Nuclear War," in Allison, Carnesale, and Nye, *Hawks, Doves, and Owls,* p. 74, suggests the utility of distinguishing between alerts for defense and retaliation and those aimed at preparing to deliver a first strike. At the same time, however, he doubts the feasibility of functional differentiation of this kind because of the similarity of the physical capabilities essential to either kind of mission.

45. For technical assessments of the feasibility of BMD, see Union of Concerned Scientists, *The Fallacy of Star Wars* (New York: Vintage, 1984); Sidney D. Drell, Philip J. Farley, and David Holloway, *The Reagan Strategic Defense Initiative: A Technical, Political, and Arms Control Assessment* (Stanford, Calif.: Center for International Security and Arms Control, Stanford University, 1984); Ashton B. Carter, *Directed Energy Missile Defense in Space* (Washington, D.C.: Office of Technology Assessment, 1984); *Ballistic Missile Defense Technologies* (Washington, D.C.: Office of Technology Assessment, 1985); and Jeffrey Boutwell and Donald Hafner, eds., *Weapons in Space* (New York: Norton, 1985).

46. Desmond Ball, *Can Nuclear War Be Controlled?* Adelphi Paper no. 169 (London: International Institute of Strategic Studies, 1981); John D. Steinbruner, "Nuclear Decapitation," *Foreign Policy* no. 45 (Winter 1981–82), pp. 16–28; Paul Bracken, *The Command and Control of Nuclear Forces* (New Haven: Yale University Press, 1983); and Blair, *Strategic Command and Control*, pp. 282–87.

47. For empirical evidence, see Lebow, *Between Peace and War*, and Robert Jervis, Richard Ned Lebow, and Janice Gross Stein, *Psychology and Deterrence* (Baltimore: Johns Hopkins University Press, 1985).

Index

Library of Congress Cataloging-in-Publication Data

Lebow, Richard Ned.
Nuclear crisis management.

(Cornell studies in security affairs)
Includes index.
1. Nuclear crisis control—United States. 2. Nuclear crisis control—Soviet Union. I. Title. II. Series.
JX1974.8.L43 1987 355'.0217'0973 86-16767
ISBN 0-8014-1989-1 (alk. paper)